SEDIMENTATION VELOCITY ANALYTICAL ULTRACENTRIFUGATION

Discrete Species and Size-Distributions of Macromolecules and Particles

SEDIMENTATION VELOCITY ANALYTICAL ULTRACENTRIFUGATION

Discrete Species and Size-Distributions of Macromolecules and Particles

Peter Schuck

Dynamics of Macromolecular Assembly Section
Laboratory of Cellular Imaging and Macromolecular Biophysics
National Institute of Biomedical Imaging and Bioengineering
National Institutes of Health, Bethesda, Maryland, USA

CRC Press
Taylor & Francis Group
Boca Raton London New York

CRC Press is an imprint of the
Taylor & Francis Group, an **informa** business

CRC Press
Taylor & Francis Group
6000 Broken Sound Parkway NW, Suite 300
Boca Raton, FL 33487-2742

First issued in paperback 2019

CRC Press is an imprint of Taylor & Francis Group, an Informa business

ISBN-13: 978-1-4987-6894-8 (hbk)
ISBN-13: 978-0-367-87828-3 (pbk)

Visit the Taylor & Francis Web site at
http://www.taylorandfrancis.com

and the CRC Press Web site at
http://www.crcpress.com

To my wife, the sunshine of my life, to whom I owe everything.

Contents

Foreword

The analytical ultracentrifuge is a very versatile instrument. It played a key role in establishing biological concepts that we take for granted today. Theodor ("The") Svedberg at Uppsala Univ. designed and built the oil-turbine ultracentrifuge[1] in the 1920s with the aim of learning whether proteins are colloids (aggregates of smaller molecules) or true macromolecules. Tanford and Reynolds tell the story in their book.[2] Svedberg began his ultracentrifuge studies with the belief that proteins are colloids, but when in 1926 he saw the single sedimenting boundary produced by hemoglobin (clearly a giant molecule), he knew he was wrong. The oil-turbine ultracentrifuge became the indispensable instrument of the protein chemist until it was replaced by the more convenient Spinco direct-drive ultracentrifuge. Schachman's book[3] greatly helped the new research field of protein ultracentrifugation emerge.

In 1958 the Meselson–Stahl experiment[4] produced another remarkable ultracentrifuge result, this time in the new field of molecular biology. The Watson–Crick DNA structure was generally accepted when it was first published in 1953. But it raised puzzling questions, as Delbruck & Stent[5] recognized. How is the DNA double helix replicated, and then separated into two daughter DNA helices, if the complementary DNA strands are continuous and wound around each other thousands of times? A new ultracentrifuge technique proved able to distinguish between semi-conservative and conservative DNA replication, although additional confirmatory experiments were still needed after the ultracentrifuge gave its answer. In the Meselson–Stahl experiment, the complementary strands of the parental *E. coli* DNA were density labeled with ^{15}N and then the *E. coli* growth medium was changed to ^{14}N and, after one generation of growth in the new medium, the two daughter DNA helices could be separated by their buoyant densities in a CsCl density gradient. The experimental results showed that both daughter DNA double helices contained equal amounts of parental (^{15}N) and daughter (^{14}N) DNA, as expected for the semi-conservative DNA replication mechanism proposed by Watson

---

[1]T. Svedberg and K.O. Pedersen, *The Ultracentrifuge*, Oxford University Press, London, 1940.

[2]C. Tanford and J. Reynolds, *Nature's Robots: A History of Proteins*, Oxford University Press, Oxford, 2001.

[3]H.K. Schachman, *Ultracentrifugation in Biochemistry*, Academic Press, New York, 1959.

[4]M. Meselson and F.W. Stahl, The replication of DNA in Escherichia coli, *Proc. Natl. Acad. Sci. USA*, vol. 44, pp. 671–682, 1958, doi:10.1073/pnas.44.7.671.

[5]M. Delbruck and G.S. Stent, On the mechanism of DNA replication, pp 699-736, in *The Chemical Basis of Heredity*, ed. W.D. McElroy and B. Glass, Johns Hopkins Press, Baltimore, 1957.

and Crick.[6] There was keen interest at that time in the alternative mechanism of conservative DNA replication (see Delbruck & Stent[5]), in which the parental DNA helix remains intact after DNA replication. For most scientists, the Meselson–Stahl experiment settled the question.

Books like this one by Dr. Schuck and his colleagues keep the methods of analyzing ultracentrifuge results up-to-date as well as taking account of any changes in basic physical concepts that underlie the methods. Thus, for the reader who is using an unfamiliar analytical method, or is deciding which method to use, it is essential to have books like this one by Dr. Schuck and his colleagues.

It was not always common to have an ultracentrifuge easily available. When I was a graduate student (1950-53), the Spinco direct-drive ultracentrifuge was not yet generally available and there were only 8 oil-turbine ultracentrifuges in the world. I began ultracentrifuge studies with an oil-turbine at the Univ. of Wisconsin with J.W. ("Jack") Williams and then did my doctoral work with another oil-turbine instrument at the Univ. of Oxford, with A.G. ("Sandy") Ogston as my adviser. The oil-turbine ultracentrifuge was a great instrument and it produced beautiful data, but for someone like myself who was not mechanically adept, it was nerve-wracking to use.

The Svedberg, who foresaw the important role that a high-speed ultracentrifuge could play in protein research, continued to design and build new rotors for the oil-turbine ultracentrifuge until he was satisfied he had reached the highest possible operating speed. According to legend, he found out how fast a steel rotor could spin by raising its speed of rotation by remote control until the rotor exploded. This took place in a shack in the woods, distant from his laboratory; he used a seismograph to record the explosion. Svedberg was a visiting professor at the Univ. of Wisconsin in 1923, when he interested Profs. Max Mason (Physics) and Warren Weaver (Mathematics) in solving the differential equation of the ultracentrifuge to give the time course of a sedimentation velocity experiment (combined sedimentation and diffusion) for molecules in the size range of proteins. (Their solution was obtained for the approximate case of constant field strength and a rectangular cell. For references to later work on the exact case (radial field and sector-shaped cell), see [7].) Weaver later became a Director of the Rockefeller Foundation and he started a new program that gave research grants in molecular biology. In his case, the ultracentrifuge led to molecular biology and the Meselson-Stahl experiment must surely have fascinated him.

Robert Baldwin, Stanford University

---

[6]J.D. Watson and F.H.C. Crick, Genetical implications of the structure of deoxyribonucleic acid, *Nature* vol 171(4361), pp 964-967.

[7]K.E. Van Holde and R.L. Baldwin, Rapid attainment of sedimentation equilibrium, *J. Phys. Chem.* vol 62, pp 734-743, 1958.

Preface

The power and beauty of sedimentation velocity analytical ultracentrifugation rests in its simplicity — the analysis of the dynamic response of dissolved particles to the application of a force that acts directly on the mass, mediated by a high gravitational field. It offers a unique window to study properties including mass, size, shape, density, composition, charge, solvation, hydrodynamics, inter-particle forces, and polydispersity. It is a classical technique with applications in virtually any scientific discipline dedicated to the physical state of dissolved particles within sizes between several Ångstrom to several micrometers and within 10^2–10^{10} Dalton, including fields such as cell biology, molecular biology, biochemistry, immunology, biophysics, physical chemistry, colloid chemistry, biotechnology, biomaterials, nanoparticles, and others.

The fundamental physical principles of sedimentation velocity analytical ultracentrifugation have been thoroughly examined and laid out by the scientific giants who followed the pioneering work of Svedberg and Pederson that led to the Nobel Prize 1926. These are timeless guideposts for the application and further development of the technique. Yet, the practice of sedimentation velocity has fundamentally changed since the monumental monographs on mathematical theory and data analysis by Hiroshi Fujita in 1962 and 1975 summarized the state of the art. In addition to major shifts in the interests of mainstream applications, improved and computerized instrumentation, and new optical detection systems with improved sensitivity allowing studies in dilute solution often without considering concentration-dependence of sedimentation parameters, perhaps the most important development is that we are now able to solve the Lamm equation — the master equation of ultracentrifugal sedimentation — accurately and rapidly enough to use it effectively in routine sedimentation velocity data analysis.

Throughout most of the 20th century, efficiently solving the Lamm equation had been a key hurdle in sedimentation velocity, due to the lack of a closed-form analytical solution of the sedimentation/diffusion process in radial coordinates. This problem constituted a major subject of theoretical mathematical development, and has dictated quantitative data analysis and significantly constrained experimental configurations for many decades. The removal of this limitation through modern computational capabilities and efficient and accurate numerical algorithms has spawned many new experimental strategies and a host of new applications. Together with modern mathematical techniques for calculating distributions drawn from other areas, such as image deconvolution and tomographic image reconstruction, the limits of sedimentation velocity data analysis have been completely redrawn: Besides the incorporation of concentration profiles at any stage of the sedimentation process, the completely new and detailed analysis of distributions, the freedom from the paradigm of a fixed rotor speed, among other extensions, we can now incorporate into the data analysis to great benefit even aspects of the data acquisition. The latter had previously been off-limits, but allows us now to create more accurate and significantly richer models, such as in global multi-wavelength analysis of multi-component samples. Most recently, even photophysical modulation of signals could be folded into the sedimentation analysis to create a new modus of multi-component detection.

The goal of this book is to describe a modern conceptual and mathematical framework of sedimentation velocity analysis general enough to naturally incorporate these developments, while detailed enough to weave together the different methods from the ground up. In this process, unfortunately, a treasure of ingenious and insightful data analysis approaches that have been developed throughout the last century, many of which have fallen out of mainstream practice, could not be included. Likewise, some side aspects of traditional methods have moved into the foreground in the current context. However, the roots of the approaches covered are highlighted by providing key references to the original literature. Hopefully, the selection of topics and references will allow the reader to absorb the creative spirit that has permeated this field for many decades, and at the same time offer a solid foundation for rigorous future developments.

The focus of the present work is limited to dissolved particles that are exposed to the centrifugal fields but do not exhibit inter-particle interactions. This includes the most common applications of analytical ultracentrifugation to date in structural biology, immunology, biotechnology, nanoscience, and colloid chemistry. The behavior of non-interacting systems is also the foundation for the analysis of particles with attractive or repulsive interactions, the topic for a forthcoming volume. Also, the present description excludes most experimental questions, and instead builds on the first volume in this series *Basic Principles of Analytical Ultracentrifugation* (referred to as Part I), which comprehensively reviews the physics of sedimentation, and leads the reader through the many aspects of how to design and carry out well-defined, meaningful experiments producing accurate data sets.

The presentation is guided by increasing physical and mathematical complexity of the sedimentation models rather than historical developments. Theoretical relationships, numerical aspects of data fitting, and experimental information content of noisy signals are presented side-by-side, as they are closely intertwined in modern sedimentation velocity analysis. Starting with the basic concepts of data fitting, the phenomenology of sedimentation velocity of single species for different experimental configurations is discussed next, which leads to the presentation of the key problem of distribution analysis: how migration from diffusion and differential sedimentation can be distinguished. This sets the stage for the systematic discussion of current strategies for data analysis with explicit distribution models in one or more dimensions, and also provides a rigorous framework to discuss some more traditional or historic analysis approaches.

The book is not thought to be a theoretical compendium, but an attempt to strike a balance between the indispensable rigorous mathematical backbone, a more intuitive physical picture of the sedimentation process, and the practical goals to highlight informative features of ultracentrifugal signal evolution and how to exploit them optimally. Therefore, a final chapter is dedicated to practical questions of data fitting. In fact, much of the material originates from a lecture series on the theory and practice of sedimentation velocity developed for our annual workshop on analytical ultracentrifugation and related biophysical disciplines at the National Institutes of Health. To further facilitate the practical application, I have highlighted, in the text, functions of the public domain software SEDFIT and SEDPHAT that carry out specific calculations and analysis tasks. The software can be freely obtained from the Dynamics of Macromolecular Assembly Section of the National Institute of Biomedical Imaging and Bioengineering at sedfitsedphat.nibib.nih.gov.

I hope the book will be a useful description of the state of the art of sedimentation velocity analysis, both for providing a modern background for experienced practitioners of sedimentation velocity analytical ultracentrifugation, as well as a systematic foundation for interested readers new to this technique. My own introduction to analytical ultracentrifugation occurred at a time of relative hiatus of the technique, when very few laboratories were knowledgeable and actively practiced it. At the time, learning this seemingly obscure technique required an apprenticeship in one of the remaining laboratories, and personally I am very grateful for being taught by Dieter Schubert, Allen Minton, Marc Lewis, and many other eminent scientists kindly sharing their knowledge. However, times have changed considerably, and with analytical ultracentrifugation becoming an essential mainstream technique once more with a steadily growing range of applications, it is important that it be possible to learn independently and read up on all important aspects. The greatest reward will be if the present book, in conjunction with Part I and planned future volumes, will help colleagues to autonomously become fully acquainted with the rich potential and the limitations and to confidently apply sedimentation velocity analyses in their research.

Bethesda, April 2016

This work was supported by the Intramural Research Program of the National Institute of Biomedical Imaging and Bioengineering at the National Institutes of Health.

SYMBOL DESCRIPTION

α — scaling parameter for regularization

$a\,(r,t)$ — radial- and time-dependent signal

$a'\,(r,t)$ — radial- and time-dependent signal in the absence of radial- and time-dependent baseline variation

$a_\lambda\,(r,t)$ — radial- and time-dependent signal at wavelength λ

$\beta(t)$ — time-dependent baseline signal offset that is radially constant (RI noise)

b_λ — radially and temporally constant baseline at signal λ

b — bottom radius (distance from center of rotation to the distal end of the solution column)

$b(r)$ — radial-dependent baseline signal offset that is temporally constant (TI noise)

$B_\lambda(r,t)$ — general radial- and time-dependent baseline

c_0 — loading concentration at time t=0

c_p — plateau concentration

$c(s)$ — sedimentation coefficient distribution

$c^{(p)}(s)$ — sedimentation coefficient distribution with Bayesian prior

$c^{(p\delta)}(s)$ — sedimentation coefficient distribution with δ-functions as Bayesian prior

$c(s,f_r)$ — size-and-shape distribution

C_k^{tot} — total concentration in MSSV mass conservation condition

$\delta(x)$ — Dirac delta function

Δt — time interval

$\Delta_{\text{fit}}^2(\gamma)$ — sum of squared residuals as a function of a distribution $\gamma(s)$; more generally also SSR

d — optical pathlength

D — translational diffusion coefficient

D^* — apparent translational diffusion coefficient

ε_λ — molar signal increment at wavelength λ

$\Phi(x)$ — erf(x), i.e., error function $\frac{2}{\sqrt{\pi}}\int_0^x e^{-t^2}dt$

$\Phi^{-1}(y)$ — inverse error function

φ — opening angle of the sector

$\phi(r,s)$ — radial- and s-value dependence of the incident photon flux in FDS detection

f — translational friction coefficient

$f^{((sed))}$ — translational friction coefficient for sedimentation

f_0 — translational friction coefficient of the equivalent compact, smooth sphere with same mass and density as the particle

$f/f_0,\ f_r$ — frictional ratio (in long and short notation)

F_b — buoyancy force

F_f — frictional force

F_{sed} — sedimentation force

$\gamma(s)$ — generic sedimentation coefficient distribution

$G(s)$ — integral sedimentation coefficient distribution

$G^*(s)$ — apparent integral sedimentation coefficient distribution of non-diffusing particles

$g^{(1)}(\tau)$ — field autocorrelation function in dynamic light scattering

$g^*(s)$ — apparent sedimentation coefficient distribution of non-diffusing particles

$\tilde{g}_*(s)$ — approximation of $g^*(s)$ based on a finite time difference

$g^\circ(v)$ — apparent velocity distribution in linear geometry with constant force

η — solvent viscosity

η_0 — standard viscosity (of water at $20\,^{\circ}C$ in 1 atm)

$H(x)$ — Heaviside step function

h — height of the sector

j — volume flux

j_{sed} — sedimentation flux

j_{diff} — diffusion flux

κ — solvent compressibility coefficient

κ_p — bulk modulus for compression of the sedimenting particle

κ_η — coefficient for the pressure dependence of solvent viscosity

κ_{all} — cumulative pressure dependence coefficient

k — enumeration of macromolecular species or component

k_B — Boltzmann constant

λ — wavelength or generalized signal

$l.h.s.$ — left-hand side (of an equation)

$\text{ls-}g^*(s)$ — apparent sedimentation coefficient distribution of non-diffusing particles $g^*(s)$ determined by least-squares fit to step functions

m — meniscus radius (distance from the center of rotation to the proximal end of the solution column)

M — molar mass

M^* — apparent molar mass

M_b — buoyant molar mass

N — resolution of a distribution

N^* — characteristic minimal resolution of a distribution

N_A — Avogadro's number

ω — rotor angular velocity

$\Omega[\gamma(s_l)]$ — general regularization functional

ϕ' — effective partial specific volume

ϕ_k — incident photon flux in fluorescence-detected SV

$p(r)$ — radial distribution of the pressure in the solution column

$p(s)$ — prior distribution in the Bayesian regularization of sedimentation coefficient distributions

(p) — superscript to denote a non-diffusing particle

ρ — solvent density

ρ_0 — standard density (of water at $20\,^{\circ}C$ in atmospheric pressure)

r — radius (distance from the center of rotation)

r_b — radius position of the boundary midpoint

$\bar{r}$ — second moment position of the boundary

r' — in the context of the transport method, an arbitrary reference radius in the solution plateau

R — gas constant

R_0 — radius of the equivalent compact, smooth sphere with same mass and density as the particle

R_S — Stokes radius

$r.h.s.$ — right-hand side (of an equation)

rms — root mean square

$rmsd$ — root mean square deviation

σ_λ — error of data acquisition at signal λ

s — sedimentation coefficient

s_w^t — instantaneous weighted-average sedimentation coefficient

s_w — weighted-average sedimentation coefficient

$s_{\text{PB,min}}$ — sedimentation coefficient of a hypothetical particle whose boundary position is the minimum radius to be fitted in each scan

$s_{\text{PB,max}}$ — sedimentation coefficient of a hypothetical particle determining the maximum radius to be fitted in each scan

SSR — sum of squared residuals

$\mathsf{S}\left[\chi\right]$ observed signal S for given spatio-temporal evolution of concentration χ

t time

t_{scan} nominal time stamp associated with a scan

$t^{(sed)}$ effective sedimentation times

T absolute temperature

v_{scan} absolute velocity of radial scan

$\bar{v}$ partial-specific volume

v_{scan} velocity of the scanner

WSSR   weighted sum of squared residuals

χ_P^2 in the context of measures for the quality of fit, the χ^2 probability distribution

$\chi_k(r,t)$ spatio-temporal evolution of the concentration distribution of species k

$\chi_{\mathrm{nd}}(r,t)$ evolution of the sedimentation profile of an initially uniform solution of non-diffusing, inert particles

$\chi_1(r,t)$ evolution of sedimentation profile at unit loading concentration

x subscript "x" in the context of GMMA enumerating different experiments

CHAPTER 1

Basic Analysis Principles

S EDIMENTATION velocity (SV) analysis is concerned with the interpretation of the temporal evolution of the radial concentration gradients of particles in solution under the influence of a centrifugal field. This is distinct in theory and practice from the thermodynamic analysis of the final equilibrium state, which is subject to sedimentation equilibrium (SE) analysis. Both are different flavors of analytical ultracentrifugation (AUC), a technique to monitor macromolecular sedimentation in solution in real time [1].

The time course of sedimentation in an SV experiment is typically recorded as two-dimensional or three-dimensional data sets consisting of radial distributions of one or multiple spectral signals at different points in time, expressed as $a(r, t)$ or $a_\lambda(r, t)$. Sometimes higher dimensional families of such data sets are available from parallel experiments in different solution conditions, or for a range of macromolecular concentrations. Optimally taking advantage of such rich data sets — using principles resting on molecular hydrodynamics, thermodynamics, and chemical kinetics, as well as optics and photophysics to extract the molecular parameters of interest — poses a formidable data analysis problem.

Before discussing specific sedimentation models and their applications, in the present chapter we will first outline the general strategy underlying modern SV analysis. We will then recapitulate briefly the basic phenomenology of sedimentation, with the goal to establish molecular sedimentation parameters that constitute essential features of any sedimentation model, as well as their fundamental relationships. This will establish the terminology and link to the basic principles outlined in Part I of this series [1].[1]

---

[1]In particular, the terminology in the current volume builds directly on the description of the physics of the sedimenting particle in Part I, Chapter 2, the description of experimental data acquisition in Part I, Chapter 4, and assumes the sedimentation experiment to be carried out as described in Part I, Chapter 5.

1.1 CONCEPTS OF MODERN SEDIMENTATION VELOCITY DATA ANALYSIS

The fundamental principle of biophysical data analysis, as applied to most of the SV analysis in the present volume, proceeds in the following steps:[2]

1. Hypothesize a molecular model of sedimentation for the molecules (potentially) observed in the sedimentation experiment, resting on fundamental forces and molecular mechanism of transport and interactions.

2. Derive from this the spatio-temporal evolution of concentration $\chi_k(r,t)$ of all macromolecular solution components k.

3. Combine this sedimentation model with a model for optical detection, $S[\chi_k(r,t)]$, given the macromolecular concentration distributions.

4. Identify known and unknown parameters in this model, establish bounds for parameter values and, if possible, relationships between unknown parameters.

5. Fit the signal model to the experimental data, refining the unknown parameters to optimize the match between data and model.

6. Accept or reject the quality of fit: If it is acceptable, assess the information content of the data for the parameters of interest, for example, by determining confidence intervals, and if the fit is unacceptable proceed to a better hypothesis regarding sample or sedimentation process.

Each of these points will be described below in more detail.

Not surprisingly, this approach is quite different from the analysis approach in the first half of the 20th century, which was necessarily based on linearizing transformations and graphical analysis [2–4]. These were largely limited to the determination of one s-value, and often required the adaptation of experimental design to produce suitable data and/or to match the approximations embedded in the data analysis approaches. A direct fit of raw sedimentation boundary data by non-linear regression with explicit models for macromolecular sedimentation was conceptually anticipated already many decades ago [4–8], but became practical only with the availability of computational hardware in the 1990s [9–12]. It became the method of choice in routine applications of SV after combination with modern mathematical tools for distribution analysis of poly- and paucidisperse samples, diffusional deconvolution, and the possibility of incorporating explicit noise models [13–15].

The modern strategy allows a statistically optimal data analysis, naturally including all meaningful acquired data. It can be further enhanced by the art of formulating a model that incorporates all available prior knowledge into the data

---

[2]Some of these steps are typically embedded into or facilitated by analysis software, but this does not change the workflow.

analysis. In this way, significantly more detail can be extracted from the sedimentation experiment than in the traditional graphical or transformation-based methods. In turn, this leads to increased reliability and accuracy of the sedimentation coefficients, and naturally makes other parameters, such as diffusion coefficients and trace populations of certain species, accessible for measurement from sedimentation data. It also provides a rational and rigorous basis to determine the statistical uncertainty of the sought parameters. Finally, the data analysis strategy of direct fitting with explicit models provides a foundation for experiments that transcend the realm of traditional SV and allow global and multi-method analyses [16].

1.1.1 The Physical Model

Developing different physical models suitable for different experimental scenarios and their implementation in practical data analysis is a major aspect of the present book. The physical model may be subdivided into a molecular model of the sedimenting macromolecules and a model of the sedimentation process itself.

The molecular model does not require atomistic detail, but needs to comprise only parameters relevant for sedimentation, such as (effective) mass, density, translational friction properties, particle–particle and particle–solvent interactions, etc. The choice of the molecular model is usually motivated by prior knowledge of the type of particles or macromolecules under study, for example, their approximate size, purity, microheterogeneity or polydispersity, and the potential to exhibit attractive or repulsive forces. Attractive interactions are usually considered in the form of discrete binding events leading to the formation of complexes with certain life-time and binding energy. Repulsive forces are obligatory at high sample concentrations (i.e., at high fractional volume occupancy) due to steric and hydrodynamic interactions, and are often modulated by additional particle interaction potentials.

The molecular model is sometimes the result of preliminary analyses and comparisons of their quality of fit. In the absence of much prior knowledge, the most commonly used model is that of an unknown, continuous distribution of freely sedimenting and diffusing particles with different sedimentation coefficients, $c(s)$ (Section 5.4). It affords diffusional deconvolution, and can be used to extract rigorous signal weighted-average sedimentation coefficients for the system under study (Section 2.3) even for systems exhibiting attractive and, to some extent, repulsive interactions. It can reveal sample purity and polydispersity, and simultaneously highlight essential features of the sedimentation process under study. This provides a good starting point for most analyses, and can serve as a basis for more detailed models that capture more specific properties of the sample under study.

The model for the sedimentation process includes the initial distribution of macromolecules, which in some cases may not be uniform, but take the shape of lamellas or step functions from the use of synthetic boundary centerpieces that can transfer and layer liquids at the start of centrifugation. In some cases, the initial distribution may also be defined by experimental data [11, 17]. The model can also include consideration of radial-dependent solvent density, viscosity, and pressure [18, 19].

Further, the sedimentation model can incorporate a time-varying rotor speed profile. For example, in modern analyses the sedimentation model does usually account for the fact that the target rotor speed of the SV experiment cannot be achieved instantaneously, but is limited by a finite rate of rotor acceleration [20]. More generally, this may involve a slow quasi-continuous decay in rotor speed as described for the rapid achievement of sedimentation equilibrium [21], a quasi-continuous increase in rotor speed for the analysis of particles with very broad size distributions [22], or other conceivable experimental configurations.

1.1.2 The Computational Model of Sedimentation

A theoretical model for the temporal evolution of the concentration distribution $\chi_k(r, t)$ is based on the master equation for sedimentation, diffusion, and interactions in the centrifugal field, the Lamm equation(s). As discussed in more detail in Section 2.2.1, it embeds the physical model discussed above, given initial distributions, temporal rotor speed profiles, and other conditions.

The Lamm equation is a parabolic partial differential equation of the convection-diffusion type. The radial geometry and radial-dependent force prohibit a general closed-form analytical solution, and exact solutions are known only for a few special cases. This is historically important, since the inability to solve the master equation of sedimentation with the accuracy and efficiency required for practical data analysis constituted a significant impediment for SV during most of the 20th century, which has been efficiently resolved only in the last two decades. Approximate analytical solutions are available for some ideal configurations, and are highly useful to provide theoretical insights [23, 24].[3] Computationally, completely general solutions are available with numerical solutions.

Numerical solutions of the Lamm equation have significantly evolved over the last several decades,[4] and the accuracy of different algorithms and their implementation must be critically kept in mind. In most cases the accuracy will generally improve with a finer discretization of space and time, at the expense of computational time. While a review of the different approaches is outside the scope of the

---

[3]A large number of approximate solutions for many special cases have been comprehensively reviewed by Fujita [24], prior to the routine introduction of computers and non-linear regression to the data analysis. Weiss and Yphantis have embarked on pioneering computational exploration of sedimentation profiles, starting with the evaluation of analytical series approximations in rectangular geometry [25]. Later, for modeling experimental data, approximate analytical solutions were pursued, independently, by Holladay [26], Behlke [27], and Philo [10], among others. In theory, these can provide equally accurate solutions as the numerical Lamm equation solutions, but known solutions are narrowly restricted to ideal experimental configurations of single non-interacting species.

[4]Numerical computational work on solving the Lamm equation goes back to the use of differential analyzers in the Yphantis lab in the 1950s [28], and high-speed digital computation has been used since the 1960s in several laboratories including those of Cox [7] and Dishon, Weiss, and Yphantis [29–33].

present work, a detailed description of the numerical approach taken in the most recent adaptive variable grid method [34] is provided in Appendix A. It allows the efficient solution of the Lamm equation with stringent error control, automatically adjusting the discretization to achieve a pre-set maximal error across the entire radial range, usually set at 0.1% of the loading concentration (or loading signal, respectively). Beyond the accuracy no further knowledge of computational details of Lamm equation solution algorithms is necessary for any conceptual or practical aspect of SV data analysis.

Access to the algorithm, discretization parameters, and desired precision of Lamm equation solutions is available in the Options ▷ Lamm Equation Options function of SEDPHAT and the Options ▷ Fitting Options ▷ Lamm Equation Parameters function (keyboard shortcut ALT-G) in SEDFIT. Defaults are set so that conservative error limits are honored, and modification of settings is usually not required.

1.1.3   Models for Optical Detection

Historically, optical detection was merely treated as a question of suitable optical engineering for designing hardware that produces a signal proportional to the local sample concentration, and as a vehicle for producing signals in different concentration ranges and different selectivity. With regard to modeling of data, properties of optical detection have, for a long time, been considered more of a nuisance, responsible for various baseline terms.

However, recent multi-wavelength and multi-signal approaches [35–37] have moved this aspect more into the foreground in SV. Also, it is increasingly appreciated that detection details beyond simple baselines need to be carefully considered for accurate modeling of experimental data [38]. This has become more involved with fluorescence optical detection [39]. Finally, with the advent of spatially and temporally modulated signals using photoswitchable fluorophores [40,41], it has become clear that the optical detection mode can generate rich additional information on the sedimentation process.

Therefore, properties of the optical detection system are explicitly considered a part of the data analysis model. In general, any combination of the following signal functions may be applied for modeling experimental data.

1.1.3.1   Signal Increments

Signal increments describe the proportionality constant between local concentration and the signal.[5] They may represent absorbance extinction coefficients, fringe shift coefficients or refractive index increments, specific molar count rates, or similar. Signal increments for the different optical systems were discussed in detail in Part I,

---

[5]Nonlinear signals (see Part I Chapter 4.3.3) may be considered, in principle, but have not been applied in practice and are therefore excluded from the present discussion.

Section 4.3. We will express them equally as ε_λ, similar to an absorbance extinction coefficient at a specific wavelength λ, but the notation is used more generally for any detection method. If a component extinction coefficient depends on the particular chemical state, such as bound/unbound, then in the present framework these states will be counted as different species with their own signal coefficient.[6]

It is useful to separate the specific signal increments from the effects of the optical pathlength, d. Therefore, in the simplest form, the evolution of the signal λ of a species k will depend on the macromolecular concentration χ_k as

$$S_{\lambda,k}\left[\chi_k(r,t)\right] = \varepsilon_\lambda d\chi_k(r,t) \tag{1.1}$$

with S denoting the measured signal. This may seem trivial, but creates the powerful framework for multi-component, multi-signal detection, which will be discussed in detail in Section 7.1.

It has been recognized that the signal increments in fluorescence optical detection may not be constant, but instead exhibit temporal and spatial gradients

$$\varepsilon_\lambda = \varepsilon_\lambda(r,t) \,. \tag{1.2}$$

Specifically, small radial signal magnification gradients often arise from a slight mismatch of the plane of rotation and scan axis [39][7]; temporal signal magnification changes occur from slow drifts in the laser intensity [39]. In first approximation, this may be modeled as

$$S_{\lambda,k}\left[\chi_k(r,t)\right] = \varepsilon_0 \times (1 + \alpha t) \times \left\{1 + \frac{d\varepsilon}{dr}(r - r_0)\right\} \times \chi_k(r,t) \,, \tag{1.3}$$

where the signal increment consists of an "intrinsic" signal increment ε_0 (dependent on excitation power, absorption and fluorescence quantum yield, overall detection efficiency, etc.) modulated by a temporal intensity drift α and a spatial gradient $d\varepsilon/dr$ [39].

---

[6]As discussed in Part I Chapter 4.3, this may occur in absorbance optical detection in the presence of hyper- or hypochromicity, or in fluorescence detection due to quenching or FRET. Non-constant signal increments may also be encountered, at least theoretically, in the interference optical detection when sedimenting buffer co-solutes modulate the refractive index contrast of the macromolecule of interest. This should be negligible in most cases, but is measurable in studies with macromolecules that are nearly refractive-index-matched to water [42].

[7]This may be (at least partially) compensated for by a choice of low focal depth, such that the detection volume remains entirely inside the sample volume [43]. However, this strategy will not perfectly remove this factor, and has the drawbacks of promoting inner filter effects [39] and exacerbation of the artifacts near the bottom of the centerpiece caused by the partial shadow of the excitation and detection cones.

SEDFIT – Temporal and radial signal magnification gradients can be switched on in the fluorescence data analysis modules, which may be invoked with the keyboard shortcut ALT-F. In SEDPHAT these fluorescence parameters are available in the monochromatic multi-component detection module [41].

Time-dependent signal increments, far beyond these simple effects of drifts in detection efficiency Eq. (1.3), occur when the intrinsic signal increment ε_0 itself is subject to temporal changes from photophysical effects. The most familiar example is irreversible photobleaching in fluorescence, but photobleaching fortunately does not usually occur at the excitation power densities used in the current commercial fluorescence detection system [40] (due to the short illumination time of samples in the spinning rotor). However, photoswitchable molecules designed for PALM super-resolution microscopy [44] can be reversibly converted into a dark state by the excitation laser during scanning in the fluorescence detection system of the analytical ultracentrifuge, or it can be converted from a dark state into a fluorescent state [40, 41]. Due to the popularity of PALM microscopy,[8] constructs of proteins of interest fused to suitable photoswitchable fluorescent proteins, such as Dronpa, are readily available [45]. Transitions of photoswitchable fluorophores affect their absorbance and fluorescence properties alike, and changes in the refractive index increment of the interference optical system are, at least in theory, unavoidable.

In the simplest case, the signal increment will follow an exponential approach toward a steady state

$$S_k\left[\chi_k(r,t)\right] = \varepsilon_0\left(1 + \beta e^{-\kappa\phi_k t}\right) d\chi_k(r,t) \qquad (1.4)$$

with a time constant κ, and amplitude factor β, which may decrease $(-1 < \beta < 0)$ or increase $(\beta > 0)$ the fluorescence quantum yield. The time constant will also depend on the incident photon flux ϕ_k

$$\phi_k = \phi_k(r, s_k) = \phi_0 \frac{2\pi\delta_{\text{beam}}}{m\omega^2 s}\left(1 - \frac{m}{r}\right), \qquad (1.5)$$

which is proportional to the laser power density ϕ_0 and the beam width δ_{beam}. A geometric correction is necessary to factor in the transition time of the molecule through a fixed-width beam at different radii, and that the total photon count incident on a molecule will therefore be slightly dependent on its position history, i.e., velocity $\omega^2 s$ (where ω is the rotor angular velocity) [41] (Part I, Section 4.3.2.3).

It is further possible to modulate the laser power density, e.g., to switch it on and off in certain intervals. In the absence of excitation light, the fluorophores tend to re-establish their initial equilibrium of photophysical states, with a relaxation constant dependent on the fluorophore. This can be further enhanced by brief illumination with a second light source at a different wavelength, often 405 nm, which

---

[8]The Nobel Prize in Chemistry 2014 was awarded for the development of super-resolution microscopy to Eric Betzig, Stefan Hell, and William Moerner.

reverses photoswitching induced by the 488-nm excitation beam in fluorescence detection. Finally, spatially non-uniform illumination applied for a brief period of time can create spatially well-defined signal features, causing additional signal boundaries specific to the photoswitchable species [41]. The main purpose of creating these chromophore-specific spatio-temporal modulation of signal increments is multi-component detection. As will be described in more detail in Section 7.2, the characteristic signal changes establish a new temporal dimension for the discrimination of components, for which, unlike multi-wavelength analysis, a single detection wavelength suffices.

1.1.3.2 Baselines

Baseline offsets are additive signal contributions not originating from sedimentation. In the most general form

$$S_{\lambda,k}\left[\chi_k(r,t)\right] = \varepsilon_\lambda d\chi_k(r,t) + B_\lambda(r,t)\,. \tag{1.6}$$

The question arises how they may be distinguished from macromolecular sedimentation signals. This requires additional constraints embedded in the optical detection method, which are described in detail in Part I, Chapter 4, and summarized in the following.

For example, the baseline may simply be a constant offset b_λ

$$S_{\lambda,k}\left[\chi_k(r,t)\right] = \varepsilon_\lambda d\chi_k(r,t) + b_\lambda \tag{1.7}$$

as may be the case, for example, in absorbance optical detection when buffer components absorb (Part I, Chapter 3), or in fluorescence detection considering the dark count or solvent Raman scattering [39].

However, even in absorbance optics, a detailed inspection of data shows that the baseline typically fluctuates locally, due to imperfections in the optical windows. This can be expressed as a radial-dependent baseline $b_\lambda(r)$

$$S_{\lambda,k}\left[\chi_k(r,t)\right] = \varepsilon_\lambda d\chi_k(r,t) + b_\lambda(r)\,, \tag{1.8}$$

also referred to as TI noise (time-invariant noise) [15, 46]. TI noise is an essential component of interference optical data, due to their exquisite sensitivity to optical pathlength differences (Part I, Section 4.1.4). There is so far no strong evidence of this noise contributing to fluorescence optical data.

Finally, a signal component consisting of a radially constant but temporally changing offset $\beta_\lambda(t)$ is often referred to as RI noise (radial-invariant noise). Along with TI noise, RI noise is an essential ingredient for any model of interference optical data. The decomposition of baseline signals in orthogonal radially constant and temporally constant components

$$S_{\lambda,k}\left[\chi_k(r,t)\right] = \varepsilon_\lambda d\chi_k(r,t) + b_\lambda(r) + \beta_\lambda(t) \tag{1.9}$$

usually does not correlate with macromolecular sedimentation signals that change simultaneously in time and space [15, 46].

More complex signal offsets may be considered if they result from a sedimentation process, for example, from sedimenting buffer components in the sample or reference sector. While these could be considered "baseline," they must be explicitly modeled [47](Part I, Section 4.3.4).

1.1.3.3 Effects of Finite Optical Resolution

The optical resolution of the detection system will depend on the focal spot size as well as the light path through the sample. The optical resolution is typically lower than the radial resolution of data points in the scans. This leads to a signal reported at a point r that arises from concentrations in an entire region σ nearby. The precise form will depend on the details of the optical system. In some cases, it may be approximated as a Gaussian convolution with width σ [39]

$$S_k\left[\chi_k(r,t)\right] = \frac{\int e^{-(r'-r)^2/\sigma^2} \chi_k(r',t)dr'}{\int e^{-(r'-r)^2/\sigma^2} dr'}, \tag{1.10}$$

and another simple model is a box-average [38]. In theory, for the absorbance optical system, the relation between signal and concentration is non-linear when considering the finite radial resolution of the measured intensity [38], but these effects are negligible in practice [38].

Another result of finite optical resolution are shadows and end effects, once the meniscus m or bottom b of the solution column are closer than the optical resolution to the report point r. A slightly different but conceptually similar situation arises in the confocal fluorescence detection system [48], which exhibits an artifact in the region close to the bottom of the solution column caused by obstruction of the excitation and/or emission cone by the bottom of the centerpiece [49]. Due to the geometry of detection, the width of this region increases with focal depth, and is usually 0.05–0.15 cm. As was shown in [39], the characteristic signal decrease can be accounted for with a model where the overall signal intensity is diminished in proportion to an obscured segment of a circle

$$S_k\left[\chi_k(r,t)\right] = \chi_k(r,t) \times \begin{cases} \frac{1}{\pi}\arccos\left(\frac{b-r}{\delta}\right) + \frac{(b-r)}{2\pi\delta}\sqrt{1-(b-r)^2\delta^{-2}} & \text{for} \quad r > b-\delta \\ 1 & \text{else} \end{cases} \tag{1.11}$$

with circle diameter δ (the effective diameter of the detection cone) [39].

1.1.3.4 Temporal Lag of Radial Scanning

Ideally, the signal distributions should represent snapshots of the radial signal profile at a time t. This will be the case, for example, for the interference optical system where the camera images the entire solution column at once. However, this is different in the commercial absorbance and fluorescence detection systems where the detection system moves across the solution column in a finite time (Part I, Section 4.2.3). Nevertheless, scan data files currently report only a single time point for the entire scan, rather than true data acquisition times for each radial data point. When observing macromolecules under conditions where they exhibit a large sedimentation velocity, an unaccounted for time lag from the finite scan velocity can lead to significant errors in the sedimentation coefficients [38]. This can be avoided when considering that the signal incorporates a variable delay across the solution column as

$$S_k \left[\chi_k(r, t) \right] (r, t_{\text{scan}}) = \chi_k \left(r, t_{\text{scan}} + (r - r_{0,\text{scan}})/v_{\text{scan}} \right) , \qquad (1.12)$$

where t_{scan} is the nominal time point when the scan was initiated at a radius $r_{0,\text{scan}}$, and assuming the scan proceeds with a velocity v_{scan} [38].

1.1.4   Fitting

The connection between the experimental raw data and the explicit model of sedimentation and detection is made by minimizing the difference between predicted and actual data points. Ideally, this provides a mathematically founded, unambiguous, one-to-one relationship between model and raw experimental data.

Only in a few cases is it advantageous to apply a transformation to the raw data prior to fitting. These include corrections in the time stamps [50], transformations to counteract non-linear radial calibration errors [51, 52], and, for analyses of scans acquired at different rotor speeds, a radial and field transformation to compensate for shifts in the entire solution column due to rotor stretching at the different rotor speeds [21]. These corrections leave the error statistics invariant. Transformations to a linear detection scale are advantageous for creating pseudo-absorbance from experimental transmitted intensity data [53], and may potentially be useful in the analysis of other data acquired with non-linear detection conditions [39] (Part I, Section 4.3.3.3).

When the model is fit to the data, the most important parameter for the goodness-of-fit in SV is the (weighted) sum of squared residuals, abbreviated "SSR," which represents the sum over all data sets and all data points of all scans:

$$\text{SSR} = \sum_{\lambda, r, t} \sigma_\lambda^{-2} \left(a_\lambda(r, t) - S_\lambda \left[\chi(r, t) \right] \right)^2 \qquad (1.13)$$

The selection of meaningful scans is discussed in Section 8.1. The SSR is statistically rooted in the usually normally distributed noise of data acquisition, which in SV is typically constant for all data points; its magnitude is termed σ_λ in Eq. (1.13). For a fit to a single data set, σ_λ may be set to 1, and the unweighted SSR is in many ways an equivalent fit criterion to the root-mean-square deviation (rmsd) of the fit, which may be written as the optimization problem

$$\underset{\{p\}}{Min} \sum_{\lambda, r, t} \sigma_\lambda^{-2} \left(a_\lambda(r, t) - S_\lambda \left[\chi(r, t) \right] \right)^2 , \qquad (1.14)$$

where $\{p\}$ is the set of unknown, adjustable parameters.

In general, the minimization of SSR Eq. (1.14) leads to a problem requiring non-linear optimization. That means that adjustable parameters occur in mathematically more complex form in the model than mere additive terms or multiplicative factors. Non-linear regression will always require starting estimates for all parameters. During fitting, the SSR is then minimized by refining the adjustable parameters in the model. A plot of SSR *vs* parameter value for each of the fitting parameters represents a multi-dimensional surface — the "error surface" — and fitting the data corresponds to the problem of finding the lowest point of the surface. Many different algorithms have been developed in applied mathematics for such optimization. Unfortunately, when non-linear parameters are involved, there is no known method that can *guarantee* in a general way that the overall global minimum is always found; therefore, different algorithms are often applied sequentially to probe for and escape merely local minima.[9]

Fortunately, many important cases involve one or more linear parameters. This includes baselines and concentration terms for given s and D. Linear parameters are very different from non-linear ones: for linear ones the minimization of Eq. (1.14) can be directly analytically carried out. In this case, the principle of separation of linear and non-linear parameters [54] offers the greatest computational efficiency. This is at the heart of modern size distribution analysis.

In SEDFIT and SEDPHAT, non-linear parameter optimization can be achieved with simplex, Marquardt–Levenberg, and simulated annealing algorithms. The FIT command will refine non-linear parameters that have been marked to be adjustable in their checkbox in the parameter window. Non-linear parameters require initial guesses, the performance of which can be tested prior to the fit with a RUN command. Linear parameters are analytically optimized in most models and do not require initial guesses.

1.1.4.1 Accommodating Time-Invariant and Radial-Invariant Baselines

A special situation also occurs for models that comprise baseline terms of the kind in Eq. (1.8) and Eq. (1.9) describing radial-invariant and time-invariant noise. The noise parameters can be separated and partial minimization of SSR may be carried out analytically with regard to the noise parameters alone. This can be done in a general way for any form of the rest of the model [15, 46].

In Part I, Section 4.1.4, we described this strategy in detail for time-invariant, radial-dependent noise of the form in Eq. (1.8). After partial minimization of Eq. (1.14), the remaining minimization problem is restricted to the remaining set of

---

[9]For complex models it can be desirable to use a strategy to avoid local minima by sequential relaxation of fitting parameters: In this approach, certain parameters that can experimentally be easily guessed are initially fixed to prior estimates during refinement of the more uncertain parameters, until the model functions grossly match the experimental data, and all unknown parameters are finally optimized.

adjustable parameters $\{p'\}$ and is of the form

$$\underset{\{p'\}}{Min} \sum_{\lambda,r,t} \left(\left(a_\lambda(r,t) - \bar{a}_\lambda(r) \right) - \left(S_\lambda\left[\chi(r,t) \right] - \bar{S}_\lambda\left[\chi(r,t) \right](r) \right) \right)^2, \qquad (1.15)$$

where the $\bar{a}_\lambda(r)$ and $\bar{S}_\lambda(r)$ are the time averages of the data and model, respectively, at radius r. The time-invariant baseline is implicitly defined for any specific set of the remaining model parameters $\{p'\}$ as [15, 46]

$$b_\lambda(r) = \bar{a}_\lambda(r) - \bar{S}_\lambda\left[\chi(r,t) \right](r). \qquad (1.16)$$

Similarly, the radial-invariant noise $\beta_\lambda(t)$ in Eq. (1.9) can be separated by replacing Eq. (1.14) with

$$\underset{\{p'\}}{Min} \sum_{\lambda,r,t} \left(\left(a_\lambda(r,t) - \bar{a}_\lambda(t) \right) - \left(S_\lambda\left[\chi(r,t) \right] - \bar{S}_\lambda\left[\chi(r,t) \right](t) \right) \right)^2, \qquad (1.17)$$

where the $\bar{a}_\lambda(t)$ and $\bar{S}_\lambda(t)$ are the radial averages of the data and model at time t, and the radial invariant baseline is implicitly defined for any model parameters $\{p'\}$ as [15]

$$\beta_\lambda(t) = \bar{a}_\lambda(t) - \bar{S}_\lambda\left[\chi(r,t) \right](t). \qquad (1.18)$$

Finally, the combination of both radial and time-invariant baseline leads to the new minimization problem

$$\underset{\{p'\}}{Min} \sum_{\lambda,r,t} \left(\left(a_\lambda(r,t) - \bar{a}_\lambda(r) - \bar{a}_\lambda(t) + \bar{\bar{a}}_\lambda \right) \right.$$

$$\left. - \left(S_\lambda\left[\chi(r,t) \right] - \bar{S}_\lambda\left[\chi(r,t) \right](r) - \bar{S}_\lambda\left[\chi(r,t) \right](t) + \bar{\bar{S}}_\lambda\left[\chi(r,t) \right] \right) \right)^2, \qquad (1.19)$$

where $\bar{\bar{a}}_\lambda$ and $\bar{\bar{S}}_\lambda$ are the overall averages of the data and model, the radial invariant baseline is as in Eq. (1.18), and the time-invariant baseline is calculated as [15]

$$b_\lambda(r) = \left(\bar{a}_\lambda(r) - \bar{\bar{a}}_\lambda \right) - \left(\bar{S}_\lambda\left[\chi(r,t) \right](r) - \bar{\bar{S}}_\lambda\left[\chi(r,t) \right] \right) \qquad (1.20)$$

(replacing Eq. (1.16)). The resulting baselines are least-squares optimal, given the experimental data, and given any particular model with parameters $\{p'\}$.

1.1.4.2 Constraints and Regularization

Constraints are highly effective ways to further embed available prior knowledge in the data analysis. They reduce the dimensionality and/or search region for the parameters of interest and can thereby decrease parameter uncertainties.

For example, constraints can be any known upper and/or lower bounds on parameter values. This may naturally arise, for example, for maximal s-values of particles due to Stokes' law given particle mass and density.[10] Constraints may also be found that relate unknown parameters to each other. For example, the molar masses of species in a monomer-n-mer system are certainly in a ratio 1:n, even though the monomer mass may not be precisely known. In this case, the mass ratio can be implemented as a constraint to relate the unknown monomer and oligomer molar masses [55].

The constraints in the examples so far are sometimes referred to as hard constraints, since they are phrased and implemented such that they will always be strictly fulfilled. In addition, there can be soft constraints, which are only applied if the data do not have sufficient information content to resolve parameters or to contradict the constraint. Technically, they are implemented as an additional optimization term to the SSR Eq. (1.14), balanced with SSR by a statistical tolerance of the rmsd provided by F-statistics.

An example for soft constraints is regularization, i.e., the requirement for distributions to be as parsimonious as possible. Regularization constraints will be introduced in the present work in more detail in Section 3.3. It is an approach to suppress excessive error amplification in the inversion of integral equations such as in size distribution analysis [14, 56]. Usually, regularization is based on measures for parsimony based on the additional knowledge of the sample under study, for example, the presence of a few fairly discrete species, the presence of an intrinsically broad size distribution, or the similarity to a distribution previously measured or hypothesized [14, 57–59]. Similarly, using mass conservation as a regularization principle, multi-signal size distribution analysis of mixtures that are spectrally poorly resolved can be constrained to produce total component concentrations corresponding to experimentally known values [37].

1.1.4.3 Global Analysis of SV and Global Multi-Method Analysis

Another strategy to improve the veracity of the model and best-fit parameters is the addition of more experimental information through the inclusion of more data sets. In SV studies of interacting systems, this is often accomplished by global analysis of data sets acquired at different concentrations [60–64] or different wavelengths [35]. Similarly, multiple experimental data sets are required for the determination of the partial-specific volume in density contrast SV [65]. Beyond the inclusion of additional AUC data sets, it is possible to combine SV with other biophysical techniques, such as dynamic light scattering, in a global multi-method analysis (GMMA) [16, 66].

In this case, it is important to give different statistical weights to the data sets from the different techniques and optimize the weighted sum of squared residuals (WSSR) instead of Eq. (1.13):

$$WSSR = \sum_x \frac{w_x}{\sigma_x{}^2} \sum_{\lambda, r, t} \left(a_{\lambda, x}(r, t) - S_{\lambda, x}\left[\chi(r, t)\right]\right)^2 \qquad (1.21)$$

---

[10]Such upper limits may be found conveniently using a **CALCULATOR** function in **SEDFIT**.

with the index "x" denoting the data sets from each experiment [16]. If all data points were only subject to normally distributed statistical data acquisition noise σ_x with a magnitude dependent on the particular experimental data set, then WSSR with $w_x = 1$ would follow the standard χ_P^2 probability distribution. However, systematic errors invariably impact data from different techniques to a different extent, and it is useful to introduce additional weight factors w_x. With a judicial choice, these factors can ensure that all data sets will contribute information to the analysis. For example, given a total number of experimental data points N_x, the choice $w_x = N_x^{-1/2}$ can compensate for very dissimilar data set sizes to make each data set contribute equally to the fit. Similarly, higher values of w_x for data of lower loading concentrations can be useful to ensure these do not get overwhelmed by small improvements in fits to the higher concentration data sets. A discussion of the choice of weight factors and related statistical tools of GMMA can be found in [16].

1.1.4.4 Verifying the Goodness of Fit

Besides the SSR (or rmsd) of the fit, it is important that we verify that the model provides an adequate description of the data. This is a critical advantage in comparison with historic data transformation approaches, and can build confidence that the model is adequate and the experiment is well understood.

Chiefly, there should be no systematicity to the residuals. This can be examined in different ways, described in detail in Part I, Section 4.3.5. The most effective approach is visual inspection of the residuals, as an overlay from residuals to different scans, as a residual bitmap [67], or as a residuals histogram [68]. A strictly statistical runs test [69] can be a helpful criterion, but is usually far too stringent for rigorous quantitative interpretation given the systematic noise patterns in SV [68].

1.1.4.5 Hierarchical Multi-Step Analysis

In some cases, for example, when studying complex interacting mixtures, it can be desirable to deviate from the above program of directly fitting the raw experimental data in favor of a multi-stage approach: First, direct fits to the sedimentation data with more general models are carried out, and salient features are identified, for example, the presence of certain peak patterns in the $c(s)$ sedimentation coefficient distribution as a function of concentration. Second, these features are quantified, for example, in the form of weighted-average sedimentation coefficients, and fitted with appropriate detailed models embedding hypothesized chemical reaction schemes.

This separation of the hydrodynamic sedimentation model from the model for chemical reactions can eliminate the need for an overly detailed description of the raw SV data that would, for example, introduce poorly determined parameters, and/or devote much computational effort to impurities that are not of interest but significantly impact the observed data [70]. The key prerequisite of this transformation of the analysis problem from the raw data space into a derived data space is, of course, that the derived quantities are faithfully extracted, rigorously matching the requirements of the subsequent analysis.

1.1.5 Estimating Parameter Uncertainties

Whether or not the best-fit parameters are meaningful will depend on how well they are defined by the experimental data. The limits of the information content of the data are sometimes not so obvious. This is true, in particular, considering that SV models are frequently quite complex, and may contain a large number of parameters, some of which may be more or less correlated, and some may be ill-defined while others at the same time may be well determined. Furthermore, parameters generally differ in their susceptibility to experimental noise. Therefore, a rigorous error analysis to determine the confidence intervals is indispensable for any parameter of interest.

The direct modeling approach to the analysis of SV data provides the opportunity for the application of powerful statistical tools to determine confidence intervals by probing the error surface Eq. (1.14). For example, focusing on different parameters one by one, we can easily determine whether, or how much, the SSR increases for suboptimal values of a particular parameter, and to what extent the other unknown parameters may compensate. F-statistics lends itself well to quantifying the statistical threshold of SSR on a given confidence level. The value of a particular parameter of interest that cannot be compensated by adjusting the remaining parameters, and thereby forces the SSR to exceed the pre-determined threshold, constitutes one side of the confidence limits of this parameter [71, 72]. Besides the confidence intervals, the shape of the error surface, or a projection of it into lower-dimensional space, will also reveal parameter correlations.

It is also possible to utilize Monte-Carlo simulations to probe how noise in the experimental data propagates into the best-fit parameters, and thereby determine a confidence interval from the percentiles of the resulting frequency distribution of parameters.[11] Even though it does not lend itself to fitting of complex non-linear models, it can offer a very useful avenue to determine confidence intervals on distribution integrals.

> In their **Statistics** menus, **SEDFIT** and **SEDPHAT** offer extensive statistical tools for the determination of parameter confidence intervals of linear and non-linear parameters, including rigorous error contour profiling, co-variance matrix analysis, and Monte-Carlo methods.

---

[11]It should be noted that this approach requires a very large number of simulations (in the thousands) to generate precise values for percentiles of high P-values. However, these iterations can be implemented very efficiently for distribution analyses without non-linear parameters.

1.2 FUNDAMENTAL FORCES AND ESSENTIAL MACROMOLECULAR PARAMETERS FOR MODELING

The determination and interpretation of macromolecular properties can often be regarded as a separate task from the data analysis in SV. For the purpose of data analysis we may condense the description of the particle to the parameters governing its macroscopic sedimentation behavior, and use well-defined but convenient parameter scales that may be different from standard conditions. As a consequence, part of the conclusions of what these sedimentation parameters mean in the context of the internal structure of the particle or the microscopic macromolecular configuration will require an extra step of parameter standardization. This separation is not universal, but highly useful to clarify the data analysis.

1.2.1 Apparent Macromolecular Parameters and the $\bar{v}$-Scale

The parameters directly observable in SV are the sedimentation coefficient s and the diffusion coefficient D, governing the migration from sedimentation and diffusion, as reflected in the macroscopic evolution of the sedimentation boundary of each particle.

It is very convenient to carry out data analysis in a scale of experimental conditions, with transformation to standard conditions of water at 20°C deferred to a later stage of quantitative interpretation, if necessary, following

$$s_{20,w} = s_{xp} \frac{\eta_{xp}}{\eta_0} \frac{(1 - \bar{v}_0 \rho_0)}{(1 - \bar{v}_{xp} \rho_{xp})} \tag{1.22}$$

(with ρ and η the solvent density and viscosity, respectively, $\bar{v}$ the partial-specific volume, and the subscripts "xp" and "0" indicating the experimental conditions and standard conditions, respectively). For example, this is an essential step prior to molecular shape analysis.

Similarly, we can often defer the interpretation of molar mass M and macromolecular partial-specific volume $\bar{v}$ and concentrate in the SV analysis first on the buoyant molar mass M_b

$$M_b = M (1 - \bar{v}\rho) , \tag{1.23}$$

which is what governs the net sedimentation force. M_b is related to s and D through the Svedberg equation (Eq. 1.35 below), and thus does not require knowledge of either $\bar{v}$ or M. However, it is usually convenient to consider a particle mass M' given a particular $\bar{v}$-scale

$$M' = \frac{M_b}{(1 - \phi''\rho)} \tag{1.24}$$

with an *ad hoc* assumption of the partial-specific volume ϕ''. If this partial-specific volume is identical to the apparent partial-specific volume ϕ', then we obtain the partial molar mass $M' = M_a$ (see Part I, Chapter 2). If it is close, then M' will be a reasonable estimate, and given an experimental sedimentation coefficient s or diffusion coefficient D, can be used to calculate a good estimate of the translational frictional ratio f/f_0 (also apparent on the particular $\bar{v}$-scale).

The relationship between true molar mass, the definitions of the particle, partial-specific volumes, the density increment, and frictional ratio are non-trivial in the presence of co-solute, when considering hydration, and/or for poly-electrolytes with associated counter-ions. The reader is referred to Part I, Chapter 2 for a comprehensive review. If we know a physically well-defined partial-specific volume, for example, $\bar{v}_{SP}$, then we can transform M' to the true molar mass

$$M_{SP} = M' \frac{(1 - \phi'' \rho)}{(1 - \bar{v}_{SP}\rho)}. \tag{1.25}$$

Similarly, these transformations may correct initial assumptions with regard to the solvent density.[12]

SEDFIT – A function in the Options ▷ Calculator menu transforms experimental s-values to $s_{20,w}$-values. The default global $\bar{v}$ value is 0.73 mL/g, and can be changed in the Options ▷ set v*rho menu.

In SEDPHAT the conversion from experimental to standard units is implicitly executed using the experimental solvent data in the experimental parameters input box for each experiment.

While it is convenient, at first, to defer further interpretation of the sedimentation parameters and stay within the operational scales, it is important to ascertain whether the measured quantities indeed represent molecular parameters, or are merely the result of an empirical fit of the observed data. This is a different question from that of scales. For example, in general we do not accumulate broadening or sharpening of the sedimentation boundary arising from polydispersity or hydrodynamic interactions into a single apparent diffusion coefficient, because the latter would become a complicated quantity that masks other processes taking place in the sedimentation experiment. Such an apparent diffusion coefficient would cease to represent molecular characteristics only, and depend on a variety of factors including the conditions of the experiment and data analysis. Rather, our goal is to analyze and unravel such interactions and mixtures explicitly to arrive at macromolecular parameters.

On the other hand, it can be useful to consider apparent sedimentation parameters of interacting systems, such as in locally weight-average sedimentation coefficients and gradient-average diffusion coefficients for rapidly self-associating systems [73, 74], or in the effective particle model for rapidly reversible heterogeneous interactions [75, 76], to the extent that these effective parameters are derived from the intrinsic properties of the macromolecular components. Similarly, apparent

---

[12]Viscosity values are irrelevant for the determination of the molar mass in SV *via* analysis of the boundary spread.

frictional ratios measured for systems with repulsive non-ideality can be extrapolated to zero concentration in a subsequent step [77, 78].

1.2.2 Fundamental Forces and the Svedberg Equation

As a foundation for SV analysis, we briefly recapitulate the physical background of macromolecular migration, described in more detail in Part I, Chapter 1.

The sedimentation of a particle in solution is governed by the net sedimentation force F_{sed}, which is in balance with the opposing translational frictional force F_f. The net sedimentation force depends on the buoyant molar mass M_b and the centrifugal acceleration $\omega^2 r$,

$$F_{\text{sed}} = (M_b/N_A)\,\omega^2 r\,, \tag{1.26}$$

whereas the frictional force depends on the (absolute) sedimentation velocity

$$F_f = -v f^{(\text{sed})}\,, \tag{1.27}$$

with f denoting the translational friction coefficient for sedimentation. With the definition of the sedimentation coefficient as the normalized sedimentation velocity

$$s := \frac{v}{\omega^2 r}\,, \tag{1.28}$$

the balance of forces leads to

$$s = \frac{M_b}{N_A f^{(\text{sed})}}\,, \tag{1.29}$$

i.e., s is a molecular constant, measured in units of Svedberg (abbreviated S with $1\,\text{S} = 10^{-13}$ sec).[13] Here, we have marked the frictional coefficient explicitly as that relevant for sedimentation, to indicate the potential for the sedimentation process to modulate the frictional coefficient by hydrodynamic drag forces inducing shifts in average macromolecular conformation.[14] However, the required forces for a deformation of even soft worm-like chains can only be generated for very large particles sedimenting at the highest possible rotor speeds, as discussed in detail in Part I, Section 2.3.2. Therefore, the distinction between $f^{(\text{sed})}$ and the ordinary translational friction coefficient f will be dropped for most of the following.

In order to relate the friction coefficient with gross size and shape, it may be rephrased *via* the Stokes radius R_S and Stokes' law

$$f = 6\pi\eta R_S\,, \tag{1.30}$$

---

[13] A simple estimate shows that Coriolis forces will be negligible: They are of magnitude $F_{\text{Cor}} = 2m\omega v = 2ms\omega^3 r$. Thus, the ratio of Coriolis force to sedimentation force is $2s\omega$. Even at the highest rotor speed, with $\omega \approx 6.3 \times 10^3$ rad/sec, and with very large particles sedimenting with 1,000 S, $F_{\text{Cor}}/F_{\text{sed}} \approx 10^{-6}$.

[14] Sedimentation effects on the molecular friction coefficient for sedimentation may be conceivable also in systems exhibiting strong charge effects, or strong non-ideality, where alignment or lateral stratification of the solution might occur (Part I, Section 2.2.2.2).

leading to

$$s = \frac{M_b}{N_A 6\pi\eta R_S} \, . \tag{1.31}$$

Further, the friction may be compared with that of a compact solid sphere with the same mass and density as the particle, which would have a radius R_0.[15] The excess friction of the particle over the sphere is the frictional ratio f/f_0, and we can interpret the sedimentation coefficient as

$$s = \frac{M_b}{N_A 6\pi\eta R_0 (f/f_0)} \, . \tag{1.32}$$

Considering particles of given shape factor (frictional ratio) and density, since the radius R_0 is proportional to $M^{1/3}$, we obtain the scale relationship

$$M \sim ((f/f_0)s)^{3/2} \, . \tag{1.33}$$

While the sedimentation coefficient governs the oriented migration of settling in the gravitational field, a second basic observable quantity in SV is the bidirectional transport from diffusion that takes place simultaneously, and in a prototypical case, leads to broadening of the migrating sedimentation boundary. The diffusional migration likewise depends on the translational friction

$$D = k_B T / f \, , \tag{1.34}$$

with k_B denoting the Boltzmann constant and T the absolute temperature.

Combining Eqs. (1.29) and (1.34) — provided the frictional coefficient for sedimentation and diffusion are identical — we arrive at the Svedberg equation [79]:

$$\frac{s}{D} = \frac{M_b}{RT} \, , \tag{1.35}$$

with R denoting the gas constant. The Svedberg equation succinctly relates the three fundamental observables in AUC experiments, which are the sedimentation velocity, the diffusion coefficient, and the buoyant molar mass.

In the default configuration of SEDFIT, the macromolecular transport parameters are expressed via the sedimentation coefficient and molar mass, using the effective partial-specific volume and the Svedberg equation (1.35) to implicitly define the diffusion coefficient. This can be changed by toggling the menu function Options ▷ Fitting Options ▷ Fit M and s.

A function in the Options ▷ Calculator menu calculates the friction force, Stokes radii, and frictional ratios for a given molar mass, apparent partial-specific volume, and sedimentation parameters.

---

[15] This is where the choice of the partial-specific volume scale enters the numeric values.

CHAPTER 2

Sedimentation of Discrete Non-Interacting Particles

S INGLE discrete classes of particles in SV experiments generate various sources of information on their sedimentation, diffusion, and buoyant molar mass. These are systematically explored in the present chapter. The term "discrete" highlights that all particles of this class in suspension are uniform with respect to size, mass, density, and friction. They may be an ensemble of states, for example, with regard to conformation, but interconvert fast on the time-scale of sedimentation to present only an average behavior. The single discrete class of particle is contrasted by the idea of a continuous distribution of particles introduced in Chapter 3, which could be thought of as an essentially infinite number of particle classes, each with slightly different parameters.

2.1   SEDIMENTATION BOUNDARIES OF NON-DIFFUSING PARTICLES

The definition of the sedimentation coefficient Eq. (1.28) can be taken as a differential equation

$$dr^{(p)}/dt = s\omega^2 r^{(p)} \tag{2.1}$$

for a point particle. If it is initially at position $r^{(p)}(0)$, its trajectory follows

$$r^{(p)}(t) = r^{(p)}(0)\, e^{s\omega^2 t}. \tag{2.2}$$

Let us imagine that at the start of sedimentation, the solution column is filled uniformly from meniscus to bottom with such non-diffusing, non-reacting point particles. From all particles in solution, those that start their sedimentation at the meniscus remain closest to the center of rotation throughout the experiment (compare Fig. 2.1 top and bottom). Therefore, a sharp boundary will form at

$$r_b(t) = m\, e^{s\omega^2 t}, \tag{2.3}$$

separating the region free of particles from that with a constant concentration of particles. Tracking the evolution of this transition point with time reveals the

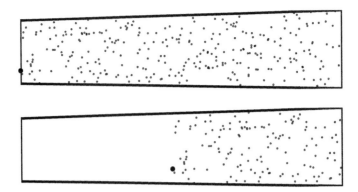

Figure 2.1 Schematic of sedimenting point particles randomly distributed throughout the solution column at the start (top) and then propagating following Eq. (2.2) for a certain period of time (bottom). One particle initially residing at the meniscus is shown as a filled bold circle.

s-value of the particle initially residing at the meniscus, m, and a plot of $log(r^{(p)})$ vs. time will be a straight line with a slope of $\omega^2 s$:

$$\frac{d}{dt}\left(log\left(\frac{r^{(p)}}{m}\right)\right) = \omega^2 s. \tag{2.4}$$

This represents perhaps the most basic way of determining s. Despite the fact that only a small fraction of all particles in the solution column are utilized to report on s — those that are initially close to the meniscus — this method will already produce remarkably good results for suitably large and homogeneous particles. However, for particles that show noticeable diffusion of the boundary, some more detailed considerations are necessary with regard to determining the proper boundary position (see below), and for smaller molecules where the migration from diffusion is comparable to that from sedimentation, the method breaks down altogether. Finally, for mixtures of particles of different s-values, curved $log(r^{(p)})$ vs. t plots might be obtained with ambiguous interpretation. Therefore, this approach is mainly of conceptual and historic relevance.

The fact that the majority of molecules will actually sediment ahead of the boundary and appear unused is interesting in two ways: First we can anticipate from here — and will see later in more detail — that in heterogeneous systems with faster and slower sedimenting species, the faster sedimenting species will always remain in a bath of the slower sedimenting ones. This is an important aspect of analyzing reacting systems that undergo dissociation/re-association events on the time-scale of sedimentation. Second, it reminds us of an alternative configuration — the method of analytical zone centrifugation described in Part I, Section 5.2.1.3 and Fig. 2.9 below, where only a lamella of protein is observed, dispensing with the seemingly little information provided by the material sedimenting ahead.

The solution plateau is not completely void of information, however, despite the fact that no gradient exists[1]: due to the exponential acceleration of the particles and the sector-shaped solution column, a dilution occurs with time that depends on the s-value of the sedimenting particles.

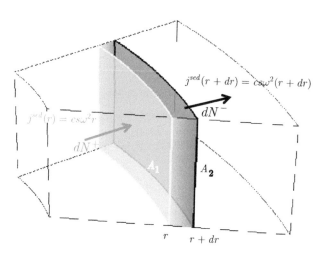

Figure 2.2 Schematics of a sector-shaped volume element at radius r with thickness dr. Per infinitesimal time interval dt, dN^+ particles enter the volume element through the inner surface A_1 (light gray), and dN^- particles leave the volume element through the outer surface A_2 (dark gray).

In Part I, Chapter 1, a derivation for the plateau concentration as a function of time was presented from within the reference frame of the sedimenting particles, based on how the inter-particle distance changes with time while they are exponentially expelled from the center of rotation. It is instructive to re-consider this result from the reference frame of fixed volume elements of the solution column by accounting for the flows into and out of a given volume element. Generally, the flux density of material is given by the product of concentration and velocity of particles,

$$ j = cv . \tag{2.5} $$

This represents the number of particles traversing a unit area per unit time. Thus, with the velocity from Eq. (1.28), the number of particles entering a centrifugal volume element at radius r during an infinitesimal time interval dt is

$$ dN^+ = A(r)c(r)s\omega^2 r dt , \tag{2.6} $$

where $A(r)$ is the area of the cylindrical sector (light surface A_1 in Fig. 2.2). During the same time interval, the number of particles leaving the volume element at the

---

[1]In a more detailed analysis, small concentration gradients in the plateau are predicted once pressure-dependent solvent density and viscosity are accounted for (Section 2.1.1). In addition, signal gradients in the plateau of ideally sedimenting homogeneous particles may arise from the finite scan velocity and/or location-dependent illumination effects in fluorescence data of photoswitchable molecules.

radius $r + dr$ is $dN^- = A(r + dr)c(r + dr)s\omega^2(r + dr)dt$ (where $A(r + dr)$ is depicted as the dark surface A_2 Fig. 2.2). The areas can be evaluated given the height h and the opening angle of the sector φ, as $A(r) = hr\varphi$ and $A(r + dr) = h(r + dr)\varphi$. Since we already know from the considerations of Part I, Chapter 1 that the concentration in the solution plateau is constant, we have $c(r) = c(r + dr) = c_p$. Combining these results, the net change in the number of particles in the volume element is $(dN^+ - dN^-) = \varphi h c_p s\omega^2 (r^2 - (r + dr)^2)$. Dropping the terms square in the infinitesimal radial increment dr, this is $-2\varphi h c_p s\omega^2 r dr$. The change in concentration is obtained by dividing this change in number of particles by the volume of the cylindrical sector, $h\varphi r dr$, leading to

$$\frac{dc_p}{dt} = -2s\omega^2 c_p, \tag{2.7}$$

which is solved by

$$c_p(t) = c_p(t_0)e^{-2s\omega^2 t}. \tag{2.8}$$

Thus, the constancy of concentration as a function of radius is in sector-shaped solution columns intimately connected with the exponential decrease of the plateau concentration with time.[2] Over the course of a typical SV experiment in a solution column of 10 mm height, the plateau concentration is reduced to about 75% of its initial value.

Inserting Eq. (2.2) for a particle initially at the meniscus into Eq. (2.8) shows the intimate link between reduction of the plateau concentration and boundary position:

$$\frac{c_p(t)}{c_p(t = 0)} = \left(\frac{m}{r^{(p)}(t)}\right)^2. \tag{2.9}$$

This is also referred to as the square dilution rule.

While in principle this drop of concentration in the solution plateau with time

---

[2]For mixtures, we obtain a form equivalent to Eq. (2.7) with the same solution Eq. (2.8) if we replace the concentration with the total concentration of the mixture and insert the (time-dependent) weighted average s-value, s_w (see Section 2.3). This can be seen the following way: For a mixture of two particle species, we can obtain the same mass balance Eq. (2.7) independently for each species. The change of the total concentration

$$c_p^{tot} = c_{p,1} + c_{p,2}$$

then follows

$$dc_p^{tot}/dt = -2\omega^2 (c_{p,1}s_1 + c_{p,2}s_2),$$

where the term in parenthesis can be rewritten via the weighted average s-value

$$s_w(t) = (c_{p,1}(t)s_1 + c_{p,2}(t)s_2)/c_p^{tot}(t),$$

leading to

$$dc_p^{tot}/dt = -2\omega^2 s_w(t)c_p^{tot}.$$

Therefore, the drop in plateau concentration with time presents an additional source of information on s_w.

could be used as a method to determine s from experimental data, by itself and in the absence of other sources of information, it would usually be a poor approach for three main reasons: First, in real experimental data, the plateau is often not perfectly flat due to contamination with faster sedimenting oligomeric aggregate species. Second, time-invariant noise offsets do not allow easily determining the plateau concentration from interference optical data, and to some extent also absorbance optical data. Third, this method would suffer from an unfavorable signal/noise ratio, measuring relatively small changes of large numbers. However, the evolution of the plateau concentration does play an important role later for the interpretation of sedimentation and chemical reactions of particles sedimenting faster than the particles forming the main boundary.

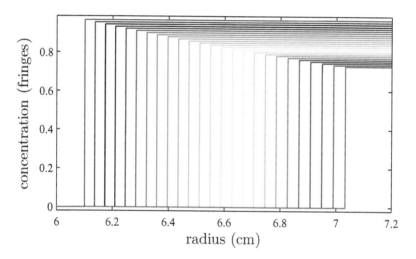

Figure 2.3 Shapes of sedimentation profiles of non-diffusing species. The data shown are simulated for a particle with 100 S sedimenting at 30,000 rpm in a 12 mm high solution column. 25 radial concentration profiles are shown in time intervals of 60 sec, with the color temperature (black to red) indicating the evolution with time. The initial time delay is from the simulated rotor acceleration at 280 rpm/sec.

Combining Eqs. (2.2) and (2.8) we can express in a closed analytical form the entire evolution of the sedimentation profile of an initially uniform solution of non-diffusing, inert particles as $\chi_{nd}(r,t)$ with the step function[3]

$$\chi_{nd}(r,t) = c_0 e^{-2s\omega^2 t} \begin{cases} 1 & \text{for } r < m e^{s\omega^2 t} \\ 0 & \text{else} \end{cases} . \qquad (2.10)$$

---

[3]With the symbols of generalized functions, this may be expressed as a Heaviside step function

$$\chi_{nd}(r,t) = c_0 e^{-2s\omega^2 t} H\left(r - m e^{s\omega^2 t}\right),$$

or with $r_b(t)$ from Eq. (2.3), $\chi_{nd}(r,t) = c_0 e^{-2s\omega^2 t} H\left(r - r_b(t)\right)$. For properties of the Heaviside step functions and the related delta-functions, see [80].

Such shapes are illustrated in Fig. 2.3. Although the step functions Eq. (2.10) are very important conceptually because they form the basis of all sedimentation profiles, for practical data analysis they are not very useful in this form. Chiefly, this is due to the presence of diffusion.[4] But even for very large, virtually non-diffusing particles, they do not account for the ubiquitous heterogeneity of the sedimenting particles. Nevertheless, the step functions can be extremely useful as basis functions for distributions, specifically the ls-$g^*(s)$ distribution introduced in Section 4.1.

The approach to derive the local concentration changes from the fluxes going in and out of a fixed volume element can be expressed in a more general way. This will be helpful for more complex scenarios where no compact analytical result can be obtained. Mass conservation imposes a general relationship between the spatial gradients of the total flux densities (in Cartesian coordinates $(\partial/\partial x)\,(j^{(\text{tot})})$; in radial geometry $(\partial/r\partial r)\,(rj^{(\text{tot})})$) and the temporal change in local concentrations $\partial c/\partial t$. This leads to a partial differential equation (PDE), which reads in the radial geometry of the centrifugal field

$$\frac{\partial \chi}{\partial t} = -\frac{1}{r}\frac{\partial}{\partial r}\left(rj^{(\text{tot})}(r)\right),\qquad(2.11)$$

with $\chi(r,t)$ denoting the radial and time-dependent concentration evolution. The only flux in the present case of non-diffusing, non-reacting particles is the sedimentation flux,

$$j^{(\text{sed})} = \chi s\omega^2 r,\qquad(2.12)$$

such that we arrive at the partial differential equation

$$\frac{\partial \chi}{\partial t} = -\frac{1}{r}\frac{\partial}{\partial r}\left(\chi s\omega^2 r^2\right),\qquad(2.13)$$

which the reader may recognize as the Lamm equation in the absence of diffusion.[5]

---

[4]One could ask at which point a hypothetical monodisperse diffusing species could theoretically be adequately described as a single non-diffusing particle. A conservative criteria would be that the rms displacement from diffusion of a real particle, $\delta_r = \sqrt{2Dt}$, not exceed the optical resolution δ_0. Combining the Stokes-Einstein law, the relationship between the Stokes radius and the sedimentation coefficient [81], and the sedimentation time from meniscus to the highest observed radius $r_{\max}$, one arrives at a condition

$$s > 150,000\,S \times \left(\ln\left(\frac{r_{\max}}{m}\right)\right)^{2/3}\left(\frac{\rho_0}{\rho_{20w}}\right)^{1/3}\frac{(\rho/\rho_0 - 1)^{1/3}}{\eta_r}\left(\frac{\delta_o}{0.01cm}\right)^{-4/3}(\#krpm)^{-4/3}$$

(where #krpm denotes the numeric value of the rotor speed in thousand rpm) for the s-value of a single sedimenting particle with density ρ forming step function boundaries in a solvent of density ρ_0 at $t = 20\,°C$. For example, for proteinaceous particles in a standard solution column with $r_{\max} = 7.1$ cm and $m = 6.0$ cm at 3000 rpm, the limit is 7000 S. However, within the finite signal/noise ratio of experimental data, distributions of non-diffusing particles will be indistinguishable from sedimentation boundaries of diffusing particles already at much lower s-values.

[5]We can show that the step function Eq. (2.10) really solves this PDE using the Heaviside step function notation for the step $H(r - me^{s\omega^2 t})$, and noting the relationship of the derivatives of H

Sedimentation models for discrete, non-diffusing particles can be invoked in SEDFIT simply by setting the diffusion coefficient to zero.

2.1.1 Non-Diffusing Species in Compressible Solvents

Due to the high gravitational fields in AUC of up to ∼300,000 g, pressures of up to ∼40 MPa can be generated at the bottom of a long solution column at the highest rotor speeds. This can have a variety of effects on the solvent and sedimenting particles, for example, in some cases modulating macromolecular interactions [82].[6] Clearly, pressure effects may be minimized by the use of shorter solution columns and lower rotor speeds than usual. However, this will often diminish the information content of the data. Therefore, it is desirable to refine the physical description of the sedimentation process to account for the basic pressure dependence of solvent density and viscosity, as well as the pressure-dependence of macromolecular buoyancy arising from macromolecular compressibility [19, 32, 83].

The weight of the solution column above different positions in the centrifugal field causes a radial increase in the pressure [19]

$$p(r) = -\frac{1}{\kappa} \log \left[1 - \frac{1}{2} \rho_0 \omega^2 \kappa \left(r^2 - m^2 \right) \right] . \tag{2.14}$$

Due to the compressibility of the solvent, this pressure gradient is accompanied by a radial dependence of solvent density $\rho(r)$, which in good approximation can be described as

$$\rho(r) = \rho_0 \left[1 + \frac{1}{2} \kappa \rho_0 \omega^2 (r^2 - m^2) \right] , \tag{2.15}$$

---

with the Dirac δ-function,

$$\left(\partial/\partial r \right) H(r - m e^{s\omega^2 t}) = \delta(r - m e^{s\omega^2 t}) ,$$

and

$$\left(\partial/\partial t \right) H(r - m e^{s\omega^2 t}) = \delta(r - m e^{s\omega^2 t}) \left(-s\omega^2 m e^{s\omega^2 t} \right) = -s\omega^2 r \delta(r - m e^{s\omega^2 t}) .$$

Thus, the time-derivative of Eq. (2.10) is

$$\frac{\partial \chi_{nd}(r,t)}{\partial t} = \frac{\partial}{\partial t} \left(c_0 e^{-2s\omega^2 t} H(r - m e^{s\omega^2 t}) \right)$$

$$= -2s\omega^2 \chi_{nd}(r,t) - c_0 e^{-2s\omega^2 t} s\omega^2 r \delta(r - m e^{s\omega^2 t}) ,$$

and the radial derivative is

$$\frac{\partial \chi_{nd}(r,t)}{\partial r} = c_0 e^{-2s\omega^2 t} \delta(r - m e^{s\omega^2 t}) .$$

Now starting from the PDE Eq. (2.13) we can expand the right hand side to

$$\frac{\partial \chi_{nd}(r,t)}{\partial t} = -s\omega^2 r \frac{\partial \chi_{nd}}{\partial r} - 2s\omega^2 \chi_{nd} ,$$

which is easily shown to be fulfilled inserting the above derivatives.

[6]For more details, see Part I, Section 2.3.1.

with ρ_0 the density at atmospheric pressure and κ denoting the solvent compressibility [19].[7] As a consequence of the pressure-induced solvent density gradient, a sedimenting particle will experience a radially increasing buoyancy force, which may be described *via* a radial-dependent sedimentation coefficient

$$s = s_0 \frac{1 - \bar{v}\rho(r)}{1 - \bar{v}\rho_0}, \tag{2.16}$$

where s_0 is the experimental sedimentation coefficient in the solvent at atmospheric pressure. This can be substituted into Eq. (2.1) for a modified differential equation of motion [19]

$$dr^{(p)}/dt = \omega^2 r^{(p)} s_0 \frac{1 - \bar{v}\rho(r)}{1 - \bar{v}\rho}. \tag{2.17}$$

With the density dependence Eq. (2.15), a particle initially at a radius r_0 will be at the position R at a later time [19]

$$R(r_0, t) = \frac{r_0 \sqrt{\Phi + \xi m^2}\ e^{\omega^2 s_0 t \frac{\Phi + \xi m^2}{\Phi}}}{\sqrt{\Phi + \xi(m^2 - r_0^2) + \xi\, r_0^2\, e^{2\omega^2 s_0 t \frac{\Phi + \xi m^2}{\Phi}}}}, \tag{2.18}$$

with the abbreviations $\Phi = 1 - \bar{v}\rho_0$ and $\xi = \bar{v}\rho_0^2\omega^2\kappa/2$. Inserting $r_0 = m$ to track the position of a particle initially at the meniscus will lead to the boundary position of a suspension of non-diffusing particles in compressible solvents. Radial integration provides an equation for the plateau concentration [19]

$$c(R(r_0), t) = c_0 \frac{\left(\Phi + \xi(m^2 - r_0^2) + \xi\, r_0^2\, e^{2\omega^2 s_0 t \frac{\Phi + \xi m^2}{\Phi}}\right)^2}{(\Phi + \xi m^2)^2\ e^{2\omega^2 s_0 t \frac{\Phi + \xi m^2}{\Phi}}} \tag{2.19}$$

at positions $R(r_0) > r_p(t)$ for an initially uniform loading concentration c_0. It defines the slightly upward sloping "plateau," describing an accumulation of particles caused by the continuous deceleration originating in the radially increasing buoyancy.

Pressure-dependent sedimentation parameters can be taken into account in SEDFIT in the Inhomogeneous Solvent function of the Options menu. This includes expressions (2.18) and (2.19) for non-diffusing species, and analogous location-dependent sedimentation coefficients in Lamm equation solutions for diffusing species. Use of these sedimentation models is compatible with distribution analyses.

Fortunately, the effect is usually quite small: with the compressibility of water

---

[7]Values for a range of solvents can be found in Part I, Appendix B. As a general rule, organic solvents often exhibit 2- to 3-fold higher values than water.

of 4.5×10^{-4}/MPa at average centrifugal pressures at 20°C only the highest rotor speeds at 60,000 rpm can create density differences in the ~1% range [19].

However, a comprehensive consideration of pressure effects should also include the compressibility of the particle, which will counteract the solvent compressibility effect, as well as the pressure-dependence of viscosity. In fact, the pressure-dependence of viscosity can be very strong in some solvents [84], easily exceeding 10% at centrifugal pressures (Part I, Appendix B). Fortunately, it is possible to additionally accommodate both the compressibility of the sedimenting particles as well as the pressure dependence of the solvent in the same quantitative framework of the equation of motion Eq. (2.17) with a pressure-dependent sedimentation coefficient [83]. With this in mind, it is useful to rephrase Eq. (2.16) directly using the approximation for the pressure underlying Eq. (2.15),

$$p(r) = \frac{1}{2}\rho_0\omega^2(r^2 - m^2), \tag{2.20}$$

such that the solvent density gradient can be written as $\rho_0\left[1 + \kappa p(r)\right]$ and the position-dependent s-value Eq. (2.16) becomes

$$s = s_0\frac{1 - \bar{v}\rho_0\left[1 + \kappa p(r)\right]}{1 - \bar{v}\rho_0} = s_0\left[1 - \kappa\frac{\bar{v}\rho_0}{1 - \bar{v}\rho_0}p(r)\right]. \tag{2.21}$$

Now generalizing this linear coefficient for the pressure-dependence of the s-value, we can introduce the compressibility of the particle itself *via* a pressure-dependent partial-specific volume

$$\bar{v}(p) = \bar{v}_0\left(1 - \frac{1}{\kappa_p}p(r)\right), \tag{2.22}$$

where κ_p is the bulk modulus of the sedimenting particle and $\bar{v}_0$ is the partial-specific volume at atmospheric pressure. Inserted in Eq. (2.21), this leads to a pressure-dependence of the sedimentation coefficient

$$s = s_0\frac{1 - \bar{v}_0\rho_0\left[1 - \kappa_p^{-1}p(r)\right]\left[1 + \kappa p(r)\right]}{1 - \bar{v}_0\rho_0}. \tag{2.23}$$

Similarly, we may define a coefficient κ_η for the pressure dependence of solvent viscosity to approximate the solution viscosity as

$$\eta(p) = \frac{\eta_0}{1 - \kappa_\eta p(r)}\bigg], \tag{2.24}$$

with η_0 the solvent viscosity at atmospheric pressure. Together, this leads to

$$s = s_0\left[1 - \kappa_\eta p(r)\right]\frac{1 - \bar{v}_0\rho_0\left[1 - \kappa_p^{-1}p(r)\right]\left[1 + \kappa p(r)\right]}{1 - \bar{v}_0\rho_0}. \tag{2.25}$$

In first approximation, dropping all terms quadratic or cubic in $p(r)$, we can rewrite as

$$s \approx s_0\left[1 - \kappa_{\text{all}}\frac{\bar{v}_0\rho_0}{1 - \bar{v}_0\rho_0}p(r)\right], \tag{2.26}$$

with the effective cumulative coefficient

$$\kappa_{\text{all}} = \kappa - \kappa_p^{-1} + \kappa_\eta \frac{1 - \bar{v}_0 \rho_0}{\bar{v}_0 \rho_0} \,. \tag{2.27}$$

Thus, in correspondence with the analysis by Fujita [83], we find that the pressure dependence of density, viscosity, and particle partial-specific volume can all be captured with the quantitative description Eqs. (2.18) and (2.19) outlined above for solvent compressibility, if we simply replace the solvent compressibility with the cumulative effective pressure dependence coefficient κ_{all} defined in Eq. (2.27).

2.1.2 Detection of Non-Diffusing Species with Finite Scan Velocity

As already indicated in Section 1.1.3.4 and discussed in detail in Part I, Section 4.2.3, the absorbance and fluorescence scanners acquire signals sequentially in the radial direction with a scan velocity v_{scan}, even though the entire scan is labeled with only one fixed time stamp.[8]

The consequences of this for theoretical sedimentation boundaries of non-diffusing particles can be readily expressed analytically as modifications to Eq. (2.10). If we accept this time stamp as a measure of sedimentation time, a correction term to the boundary movement Eq. (2.2) is necessary to account for the migration during the increasing extra time lag (or decreasing extra time lag for scanners in negative direction) before the boundary is detected [38]. Likewise, as the scanner proceeds along a plateau, the extra time required to reach a certain radius will alter the observed plateau concentration. In the notation of the signal this is

$$S\left[\chi_{nd}(r, t_{\text{scan}})\right] = c_0 e^{-2s\omega^2 \left(t_{\text{scan}} + \frac{(r - r_{\text{scan}})}{v_{\text{scan}}}\right)} \begin{cases} 1 & \text{for } r < m e^{s\omega^2 \left(t_{\text{scan}} + \frac{(r - r_{\text{scan}})}{v_{\text{scan}}}\right)} \\ 0 & \text{else} \end{cases},$$

$$\tag{2.28}$$

assuming the time stamp t_{scan} is recorded when the scanner is at the position r_{scan} [38]. When scanning boundaries of sedimenting particles from smaller to higher radii, Eq. (2.28) describes a slightly decaying exponential.

A flag can be set in the Lamm Equation Parameters in both SEDFIT and SEDPHAT to account for finite scanning velocity. There, the user can enter an experimentally determined scan velocity value.

---

[8]In the absorbance system, the velocity is ~ 2.5 cm/sec under standard SV scan settings of 0.003 cm radial increments and a single acquisition per radius. In the fluorescence system, it is ~ -1.3 cm/sec.

2.2 DIFFUSING PARTICLES AND THE LAMM EQUATION

The Lamm equation incorporates diffusion into the description of the sedimentation process. This is obviously an essential component of a realistic picture of almost all sedimentation experiments. Only particles with masses in the high MDa or GDa range and above may — under suitable experimental conditions — be approximated well with the ideal concentration profiles of non-diffusing species $\chi_{nd}(r, t)$ described above. The ability to extract both sedimentation and diffusion coefficients from an SV experiment promises information on the buoyant molar mass M_b (via the Svedberg equation $M(1 - \bar{v}\rho) \sim s/D$, Eq. (1.35)). In practice, however, this potential cannot be realized until we also have a tool at our disposal to unravel heterogeneity (Section 5), since both the ubiquitous sample heterogeneity and diffusion jointly determine the boundary spread, in addition to chemical reactions or other interactions, if present. Nevertheless, the inclusion of diffusion in the theoretical description of sedimentation is an essential foundation for all modern sedimentation analysis.

2.2.1 Phenomenology and Derivation of the Lamm Equation

In the framework for non-diffusing particles, we have discussed the relationships Eqs. (2.11)–(2.13) of how the evolution of concentration profiles can be determined by considering mass conservation imposing a general relationship between the spatial gradients of the total flux densities $(\partial/r\partial r)(rj^{tot})$ and the temporal change in local concentrations $\partial c/\partial t$. Simply put, for example, if there is less material coming out than going into a volume, there must be local accumulation reflected in an increase in local concentration.

As illustrated in Fig. 2.4, we can extend the previous considerations to include diffusion in the total flux density of particles in and out of a volume element,

$$j^{(\text{tot})} = j^{(\text{sed})} + j^{(\text{diff})}, \tag{2.29}$$

with the sedimentation term $j^{(\text{sed})} = \chi s\omega^2 r$ (from Eq. (2.12)). The diffusion fluxes are proportional to the concentration gradients according to Fick's first law $j^{(\text{diff})} = -D\partial\chi/\partial r$. Together, this gives

$$\frac{\partial \chi}{\partial t} = -\frac{1}{r}\frac{\partial}{\partial r}\left(\chi s\omega^2 r^2 - D\frac{\partial \chi}{\partial r}r\right) \tag{2.30}$$

with the boundary conditions for the ends of the solution column to be impermeable, $j^{(\text{tot})}(m) = j^{(\text{tot})}(b) = 0$. This PDE was derived first by Ole Lamm in 1929 [85]; for a more detailed derivation, see Fujita [24]. Numerical solutions to this equation are the basis for modern SV analyses, and various aspects will be discussed in the following. Unless otherwise mentioned, the initial condition is $\chi(r, 0) = c_0$ for uniform loading of the sample at start of centrifugation (for alternatives, see Section 2.2.3). Also, for simplicity, the rotor speed is conventionally assumed to be time-invariant ω_0, although realistic numeric solutions will treat the rotor speed

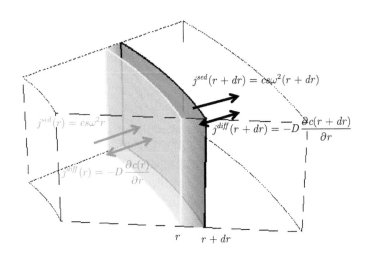

Figure 2.4 Schematics of the transport fluxes into (grey) and out of (black) a volume element. The sedimentation fluxes are unidirectional always toward the bottom of the cell, whereas the diffusion fluxes can transport material in either direction, depending on the local concentration gradient. The sum of all these fluxes determines the net gain or loss of material in the volume element per time.

as variable $\omega(t)$, at minimum to accurately describe unavoidable rotor acceleration time (Section 2.2.4).

Examples for the family of traces described by the Lamm equation for different conditions are shown in Fig. 2.5. The sedimentation patterns range from very shallow gradients relatively quickly increasing in slope (Panel A), "approach-to-equilibrium" type broad gradients that evolve toward an exponential (Panel B), classical sedimentation boundaries increasing in steepness with increasing particle mass (Panels C, D, and E), and flotation patterns that appear similar to mirror images of the sedimentation patterns except for the radial increase in concentration substituting the radial dilution (a representative case is shown in Panel F). In principle, flotation patterns may be generated corresponding to all types of sedimentation patterns shown.

In comparison to the ideal boundaries of non-diffusing species, it appears that diffusion has a very strong influence on all sedimentation processes except perhaps for those of the largest particles in Panel E. For the smallest particles (Panels A and B) strong diffusion counteracting the sedimentation is responsible for the concentration gradients spanning the entire solution column. In Panels D and E, the diffusion can be imagined to be superimposed to the movement of the sedimentation boundaries seen in the non-diffusing case, acting to broaden the sedimentation boundaries increasingly with time. Here, the diffusion can counteract the sedimentation only where the very strong concentration gradients appear in the close vicinity of the bottom (constituting the back-diffusion region). Panel C is an intermediate case

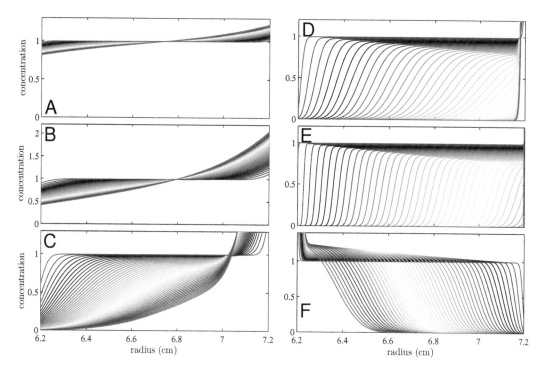

Figure 2.5   Examples for shapes of Lamm equation solutions. The profiles are calculated for particles of different size and sedimentation properties. All conditions are calculated for a rotor speed of 50,000 rpm, in standard solvent, and 50 concentration profiles are shown at different (in each case equally spaced) time intervals. Later scans are indicated by a higher color temperature. (A) Small molecules with a sedimentation coefficient of 0.2 S and a diffusion coefficient of $6 \times 10^{-6} \mathrm{cm}^2/\mathrm{sec}$, in time intervals $\Delta t = 300$ sec. Similar sedimentation parameters are frequently observed for buffer salts. (B) Sedimentation of a peptide of 1 kDa and 0.3 S, with $\Delta t = 1000$ sec. (C) A small protein of 10 kDa and 1.5 S, $\Delta t = 500$ sec. (D) A protein of 100 kDa and 6 S, $\Delta t = 300$ sec. (E) A particle of 1 MDa and 30 S, $\Delta t = 50$ sec. (F) A floating particle with a sedimentation coefficient of -3 S and a diffusion coefficient of $2.71 \times 10^{-7} \mathrm{cm}^2/\mathrm{sec}$. Such data patterns may be obtained, for example, with large emulsion or lipid particles. Note that here the radial dilution is replaced with a radial increase in concentration in the plateau region of successive profiles.

that shows moving boundaries, but they are broad and the back-diffusion region extends significantly into the solution column.

For a "classical" SV configuration (Panels D and E), the traces resemble error functions (and Gaussians for analytical zone centrifugation). This resemblance would indeed be exact if the solution column would be rectangular and the driving force would be constant like, e.g., in electrophoresis. But quantitative differences from error functions arise in SV analytical ultracentrifugation both due to the radial dilution in the sector-shaped geometry and the exponential acceleration of the particles in the centrifugal field. Over extended periods ($t \to \infty$), all Lamm equation solutions will approach a time-independent Boltzmann distribution, i.e., sedimentation equilibrium.

Lamm equation solutions are realistic models for concentration profiles to be expected experimentally, as illustrated in Fig. 2.6. However, samples following such a single species ideal sedimentation pattern are not commonly encountered due to the ubiquitous presence of at least a low level of aggregation and degradation products in biological samples and residual polydispersity in synthetic samples, respectively. Therefore, in practice, the evolving signal profiles of most samples sedimenting under suitable conditions can be described extremely well as a super-position of Lamm equations, one corresponding to each species in solution.

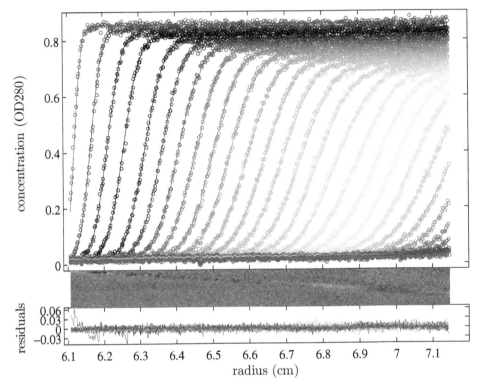

Figure 2.6   Example for a fit of experimental data with a single-species Lamm equation solution. An IgG sample in phosphate-buffered saline was purified by size-exclusion chromatography and sedimented at 50,000 rpm. The best-fit sedimentation parameters are $s = 6.286$ S and $D = 4.237$ F. The loading signal is 141-fold the rmsd of the best fit. No significant improvement in the quality of fit can be obtained with either a $c(s)$ model to account for micro-heterogeneity, or a model with repulsive hydrodynamic non-ideality. An unexplained slight systematic feature is visible in the residual bitmap at ~6.8 cm, which may be hypothesized to originate potentially from non-linear radial calibration errors [52], or from undetected deformation of the centerpiece rendering it imperfectly sector-shaped.

2.2.2   Approximate and Special Case Solutions

The Lamm equation Eq. (2.30) has no closed-form analytical solution, which has hampered many aspects of SV analysis for most of the 20$^\text{th}$ century. Efficient numerical solutions now exist, as described in Appendix A, essentially removing limitations in solving (2.30) in any form. However, despite the success of numerical solutions, their drawback is that they only provide insight into particular solutions for particular parameter values, which cannot easily be understood in a more general framework: numerical solutions do not lend themselves to derive general principles beyond the derivation of their underlying PDE. Furthermore, without theoretical reference points, it is non-obvious that a numerical solution of the PDE will actually converge correctly to the true solution — a problem exacerbated with increasing complexity of the Lamm equation model.[9]

Prior to the abundant availability of efficient computational hardware, many approximate or special case solutions of the Lamm equation have been derived over the decades [10, 23, 27, 86–89], in some cases guiding experimental design and data analysis. They can still be highly useful to highlight particular theoretical relationships.

The first analytical solution was derived by Faxén [23], in the form of a series approximation for the concentration profiles with an initial step in an otherwise infinitely long solution column, or with boundary conditions for an impermeable wall representing either a meniscus or bottom. Fujita and MacCosham have developed an approximation including boundary conditions for both the meniscus and bottom [88]. The first term may be written as

$$\chi(r,t) = c_0 e^{-2\omega^2 st} \frac{1}{2} \left[1 - \Phi \left(\frac{m \left(\omega^2 st + log(m) - log(r) \right)}{2\sqrt{Dt}} \right) \right], \qquad (2.31)$$

where Φ is the error function $(2/\sqrt{\pi}) \int_0^x e^{-t^2} dt$. As shown in Fig. 2.7, clearly the concentration profiles predicted at this level of approximation are only a semi-quantitative representation of the sedimentation boundary, but describe the sedimentation process remarkably well.[10] Therefore, Eq. (2.31) may be used as a theoretical expression for a first-order approximation of the effect of diffusion.

The radial derivative of Eq. (2.31) leads to an approximation for the central slopes of the sedimentation profiles

$$\frac{d\chi(r_b, t)}{dr} = c_p \frac{m}{r_b} \frac{1}{2\sqrt{\pi Dt}}, \qquad (2.32)$$

---

[9]An illustration of this point are numerical Lamm equation solutions based on finite differences motivated by fluxes between physically inspired radial bins. Such solutions are subject to "numerical diffusion," which is a well-known artifact in the magnitude of modelled diffusion arising from the assumption of uniform concentration within the bins [34]. However, except for the magnitude of diffusion, nothing in these solutions would indicate the errors involved.

[10]Higher-order terms in combination with additional correction terms motivated *ad hoc* for better numerical results were applied by Holladay [89], Philo [10], and Behlke and Ristau [27] for quantitative data analysis based on analytical Lamm equation solutions. Drawbacks of analytical solutions for practical data analysis are their inflexibility with regard to boundary conditions and initial conditions, and their limitations with regard to attractive or repulsive interactions, as well as consideration of time-varying rotor speeds. However, most of the material discussed in the present volume — especially the distribution analysis — could, in principle, be equally carried out using these analytical solutions, but only discrete ideal species models have been implemented in shared software such as **LAMM** and **SVEDBERG**.

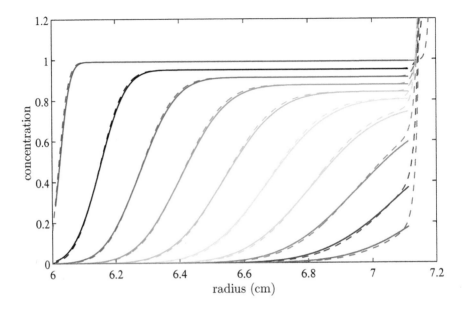

Figure 2.7 Accuracy of the first term of the Faxèn solution to the Lamm equation, calculated for a 3S, 40 kDa-protein species sedimenting at 50,000 rpm (solid lines). For comparison, the dashed lines show the correct concentration profiles from finite element numerical Lamm equation solutions. Except for the back-diffusion region, which is not a feature of the Faxèn solution, the maximum deviations are ~5% of the loading concentration and occur when the boundary has not separated yet from the meniscus.

where the boundary midpoint r_b follows Eq. (2.3) and the plateau concentration c_p follows Eq. (2.9) [38]. This shows how the information from the boundary height and position (i.e., the rate of sedimentation) together with the rate of dispersion are sufficient to determine the molar mass of a single sedimenting species.[11]

A very useful aspect of all Lamm equation solutions was identified by Archibald [90]: With the impermeability of the meniscus and bottom and the locally vanishing total flux, $j^{(\text{tot})}(m) = j^{(\text{tot})}(b) = 0$, it follows with Eq. (1.35) that at any time

$$\frac{1}{m\,\chi(m,t)}\frac{\partial\chi(m,t)}{\partial r} = \frac{1}{b\,\chi(b,t)}\frac{\partial\chi(b,t)}{\partial r} = \frac{M_b\,\omega^2}{RT}\,. \qquad (2.33)$$

Therefore, the relative slope of the concentration profiles at the meniscus and at the bottom will report the buoyant molar mass of the solute throughout the entire time of the experiment. Despite the practical limitations of directly imaging the ends of the solution columns due to unavoidable optical artifacts, at relatively low rotor speeds and early times the curvature at the extreme values is not very high, and extrapolation to the meniscus and bottom radius is straightforward.[12] Suitable conditions exist, for example, in Panel B of Fig. 2.5 and as illustrated in Fig. 2.8.

---

[11]This is closely related to the historic analysis of Schlieren data, which represent optical radial derivatives of concentration profiles, utilizing the height/area ratio of SV boundary peaks.

[12]Small errors are introduced from the uncertainty of the meniscus position [91].

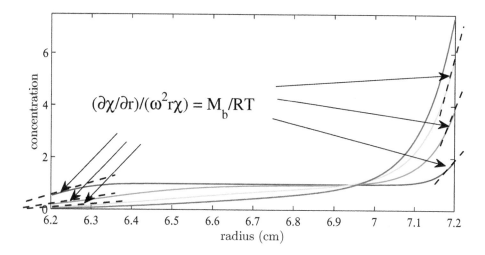

Figure 2.8 Approach to equilibrium SV concentration profiles, highlighting the local slopes at the meniscus and bottom (dashed lines), which according to the Archibald relationships Eq. (2.33) are always proportional to the product of local concentration and buoyant molar mass.

The Archibald method to measure molar masses from the slopes at the ends of the solution column is a classic approach in AUC, with the virtue of providing some equilibrium information very early in the sedimentation experiment. Even with modern Lamm equation modeling it is possible to utilize the same information content in regions close to the meniscus and bottom at early times, allowing one to rapidly determine average buoyant molar masses from approach-to-equilibrium type experiments [92].

An exception to the reflective boundary conditions of the Lamm equation, giving rise to the Archibald conditions, can exist for large particles that generate very high concentrations and steep gradients at the bottom of the cell (e.g., Panels D and E in Fig. 2.5). Under these extreme concentrations, particles are likely to exhibit strong repulsive non-ideal interactions, or may pellet and form surface films. Additionally, this region close to the bottom of the cell cannot be reliably imaged and has to be excluded from the data analysis. Correspondingly, if excluded from the analysis, the numerical solution of the Lamm equation likewise does not always need to include the back-diffusion feature. In the absence of long-range electrostatic forces, the free fall of large particles in the centrifugal field is not influenced in any way by the manner in which the particles accumulate in the region very close to the bottom. This renders this boundary condition irrelevant, providing the option to use permeable boundary conditions that make the Lamm equation much easier to compute [34, 61](Appendix A).

2.2.3 Initial Conditions

The initial condition, i.e., the concentration distribution of each species at the start of experimentation, warrants a more detailed discussion. As a law for the evolution with time, the Lamm equation itself does not specify how the molecules were distributed initially at the start of sedimentation. Rather, this is determined by the experimental setup. For multi-component systems an initial condition is necessary for each component.

A conceptually straightforward and general approach is the use of an experimental scan as the initial condition, $\chi(r, t_{scan1}) = a(r, t_{scan1})$ (see Part I, Fig. 5.4) [11,17]. It has the advantage of not requiring any reference to what happened in the experiment before this particular scan, be it convection, transient leakage, or imperfect overlaying in synthetic boundary experiments. An elegant aspect is that statistical noise in the scan serving as the initial condition does not propagate much further into the analysis due to the diffusion terms in the Lamm equation. However, a serious drawback is the difficulty of distinguishing macromolecular signals from baseline signals in the initializing single scan, which is especially problematic in the presence of TI noise. Further, an important limitation of this approach is that it can only provide a single initial condition, which usually allows application only to samples of a pure, single species or a pure self-associating component (or n components if n signals can be acquired, respectively). Such purity requirements are not often fulfilled in practice.

In order to accommodate initial conditions in sufficient number for all species considered to be potentially in solution, geometric conditions imposed by the ideal experimental design are usually applied. In the classical configuration, where the solution column is loaded uniformly at the start of centrifugation, we have the uniform initial condition $\chi(r, t = 0) = c_0$. This is the standard assumption, used in the simulated examples of Fig. 2.5 and the experiment Fig. 2.6 above. It is a very safe assumption provided the samples are well-mixed before the start of the run.

Alternatively, the initial conditions may consist of one or more step functions, if a synthetic boundary was created at the start of the experiment. For example, this may describe a lamella of a sample initially on top of a solvent column, as in band centrifugation, alternatively referred to as analytical zone centrifugation illustrated in Fig. 2.9 (see also Part I, Section 5.2.1.3). Similarly, a solvent layer may be created on top of the sample solution with a synthetic boundary centerpiece, as sketched in Fig. 2.10 (see also Part I, Section 5.2.1.2).

In contrast to the uniformly loaded samples, in these latter experiments it is less certain that the theoretical initial condition of a step function is really well approximated by the overlaying process. This is one of the main reasons why they are not as commonly applied for quantitative SV. Nevertheless, they have specific properties that can make them favorable in some cases. Briefly, for the synthetic boundary experiment, a particular strength is the measurement of diffusion coefficients of small molecules. The initial lack of signal from the molecules of interest in the upper half of the solution column and the initially high concentration gradient ensures the observation of large diffusive transport. Further, when using the interference optics it allows directly measuring the total signal of small solutes, a

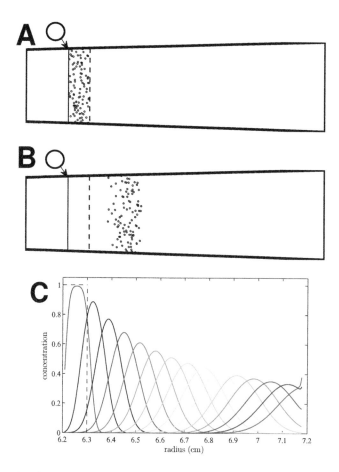

Figure 2.9   Principle of analytical zone centrifugation. *Panel A*: Schematics of an analytical zone centrifugation experiment showing the particles loaded above a column of solvent, gravitationally stabilized by a solvent composition ensuring higher density. Transfer from the reservoir (circle) is accomplished through a thin channel, which creates sufficient surface tension to keep the sample in the reservoir until a critical centrifugal field strength is achieved during the rotor acceleration phase that allows the sample to flow into the sector. The situation depicted is at a time immediately after the overlaying process is completed. *Panel B*: The evolution after a period of time, allowing the lamella of molecules to sediment and to diffuse. *Panel C*: Theoretical analytical zone centrifugation profiles predicted by the Lamm equation for a protein of 100 kDa and 6 S, shown in time intervals $\Delta t = 600$ sec, sedimenting at 50,000 rpm, with an initial lamella width of 0.1 cm (indicated by dashed black lines). The profiles resemble, but are not identical to Gaussians. Experimental data from such a configuration can be found in Part I, Fig. 5.5.

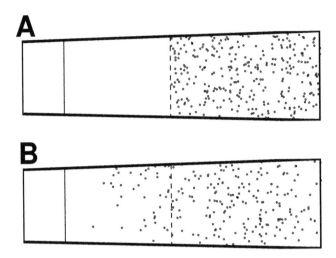

Figure 2.10 Principle of a synthetic boundary experiment. During the initial acceleration of the rotor, hydrostatic pressure develops and forces solvent from the reference sector, which has been loaded with a higher solution column, through a thin channel over the sample solution in the sample sector. This creates initially a sharp boundary between solvent and sample solution A. At later times (B), sedimentation will take place, but at the same time diffusion will transport some of the sample molecules into the region initially void of sample. An example for synthetic boundary data can be found in Part I, Fig. 5.3.

quantity that can be difficult to obtain otherwise.[13] For analytical zone centrifugation, among the specific advantages are the low sample consumption and the ability of a rapid buffer exchange.

Initial conditions of band centrifugation and synthetic boundary experiments are available in SEDFIT in the Analytical Zone Centrifugation subset of the Model menu. Synthetic boundary conditions are described with the invert band flag, available in the discrete species model phrased in the s,M-parameterization.

2.2.4 Time-Varying Centrifugal Fields

The consideration of time-varying centrifugal fields $\omega(t)$ goes back to the theoretical work of Gehatia in 1965 [93] and Nossal and Weiss in 1970 [94] on approximate

---

[13]Due to the lack of an absolute fringe displacement reference, one usually has to resort to computational analysis of the evolution of gradients with Lamm equation fitting in order to determine the total loading signal. For larger molecules that clear the meniscus at least partially, the loading concentration is a very well-determined parameter. Unfortunately, for very small solutes that exhibit only very shallow gradients (such as in Panel F of Fig. 2.5), a correlation between s-values, D-values, and the total loading concentration appears, making the latter only accessible if independent information on s- or D-values can be provided.

analytical solutions of the Lamm equation for arbitrary centrifugal field profiles. For a non-diffusing point-particle, integration of the differential equation of motion Eq. (2.7) is easily possible with an arbitrary $\omega(t)$, leading to the same trajectory Eq. (2.8) if the time t is substituted by an effective sedimentation time

$$t^{(\text{sed})}(t) = \omega(t)^{-2} \int_0^t \omega^2(t')dt' \tag{2.34}$$

with $\omega(t)$ the instantaneous rotor angular velocity at time t, where t is the true elapsed time since the start of the centrifugation.[14] However, unfortunately, the same time transformation does not hold true for diffusion [93–95]. Therefore, solution of the full Lamm equation (2.30) explicitly accounting for the time-varying centrifugal field for both sedimentation and diffusion is essential for an accurate model of SV.

In the last decade, time-varying rotor speeds have been incorporated in sedimentation models taking advantage of numerical solutions of the Lamm equation (2.30) to represent faithfully the rotor acceleration process, which takes ∼3 min for a standard rotor speed of 50,000 rpm, both in sedimentation and diffusion [20]. Beyond the initial linear rotor speed increase, it has recently become possible to numerically solve the Lamm equation Eq. (2.30) numerically for any time-varying rotor speed $\omega(t)$ [21, 22]. In parallel, it was shown that it is straightforward to experimentally carry out complex rotor acceleration and deceleration schedules in a well-defined manner, and without causing convection [21, 22]. This enables new configurations for SV that have not been previously amenable to consideration by analytical solutions of the sedimentation/diffusion process.

The implications of this have not yet been fully explored. It was recently exploited to compute a schedule for the centrifugal field that will significantly reduce the time required for attainment of sedimentation equilibrium [21] (Part I, Section 5.7.2). For SV, an application based on a concept developed by Scholtan and Lange is termed "gravitational sweep sedimentation," where a slow quasi-continuous increase of the rotor speed causes continuously smaller particles to sediment [22, 96, 97]. This has the benefit of increasing the sedimentation coefficient range of particles that can be studied in a single experiment.[15] In addition, slowly

---

[14]For a derivation see Part I, Section 4.2.1.

[15]SV experiments with continuously increasing rotor speeds have previously been employed in the study of very large, non-diffusing particles [96, 97]: Under the approximation of the effective sedimentation time, the temporal signal from a fixed-radius detector was converted into a sedimentation coefficient distribution and, in turn, transformed — using assumptions of particle density and shape — into a particle size distribution. Related, a multi-speed method combining data acquired at a few, step-wise increasing rotor speeds (going back to an unpublished method by Yphantis [98]) was used by Stafford and Braswell [99], also based on the same effective sedimentation time approximation of non-diffusing particles, for calculating combined apparent sedimentation coefficient distributions $g(s^*)$ from the time difference of scans described in Section 6.2. Analogously, a radial derivative transformation was described by Mach and Arvinte [100] for the determination of (apparent non-diffusing) particle size distributions from scans acquired at increasing rotor speeds. These methods fail in the presence of smaller particles exhibiting significant diffusion.

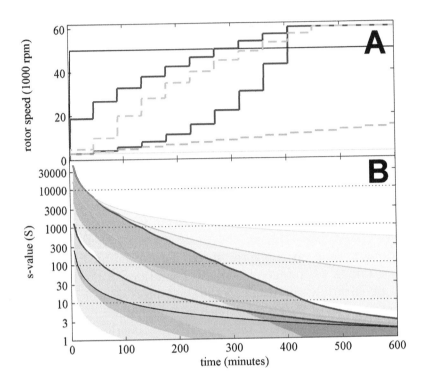

Figure 2.11   Different rotor speed profiles in gravitational sweep SV and their impact on the observable range of sedimentation coefficients throughout the experiment. *Panel A*: Rotor speed models following a constant field increase model (blue), a power-law model (red), an *ad hoc* modification of the constant field increase model extended by two lower speed steps at 5000 rpm and 10,000 rpm (green dashed), and a model with slowly linearly increasing rotor speed (orange dashed). Rotor speed changes were initiated in intervals of 45 min, with rotor acceleration of 280 rpm/sec, from a minimal rotor speed of 3000 rpm to a maximal rotor speed of up to 60,000 rpm. For comparison, constant speed data are shown for a conventional constant speed experiment at 50,000 rpm (black) and 3000 rpm (cyan). *Panel B*: Observable range of *s*-values for non-diffusing particles located between 6.1 cm and 7.1 cm in a solution column of 12 mm height, indicated with patches colored corresponding to the rotor speed schedules in *A*. Reproduced from [22].

ramping up the rotor speed can avoid adiabatic cooling of the rotor at the start of centrifugation, enabling the experiment to be carried out close to isothermal [95].

In gravitational sweep sedimentation, different rotor speed schedules $\omega(t)$ can be envisioned with different implications for the sedimentation boundary shapes and particle size range, as illustrated in Fig. 2.11 [22]. For example, the exponential model shown in red makes it possible to observe, in a single run, the entire range of particle sizes from ∼0.1 S to up to ∼100,000 S. A side-effect of initial low-speed phases is that sedimentation of smaller particles is observed later and their boundaries exhibit slightly more diffusion broadening than in the conventional constant field conditions. Calculated sedimentation boundary shapes of different size species

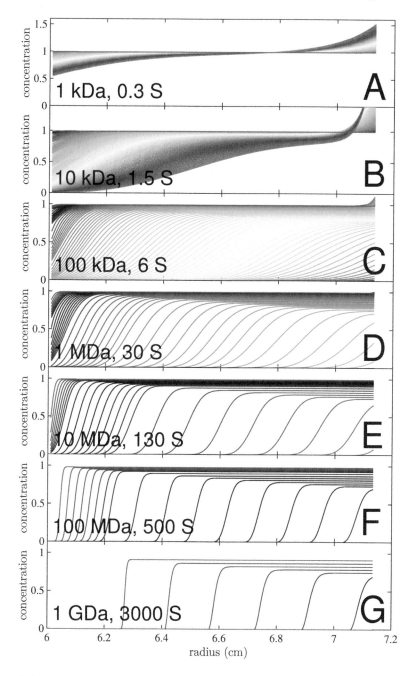

Figure 2.12 Calculated boundary profiles of a single discrete species sedimenting in variable field SV under conditions of the schedule shown in green in Fig. 2.11. Radial concentration profiles are calculated for a time-period from ~9 min to 8.5 hours, in 100 scans of 5-min intervals. All panels show the radial profiles at the same time points and using the same color scheme with increasing color temperature indicating later time. For reference, the green range is approximately in the middle of the experiment where the rotor speed is 40,000 rpm. Sedimentation was simulated for particles with partial-specific volume of 0.73 ml/g in water, with frictional ratios between 1.2 and 1.5. Sedimentation parameters are for species of (A) 1 kDa, 0.3 S; (B) 10 kDa, 1.5 S; (C) 100 kDa, 6 S; (D) 1 MDa, 30 S; (E) 10 MDa, 130 S; (F) 100 MDa, 500 S; (G) 1 GDa, 3000 S. Reproduced from [22].

are shown in Fig. 2.12 for the extended constant-field increase model (green in Fig. 2.11).

Time-varying centrifugal fields can be implemented experimentally either through pre-programmed rotor speed schedules in the "equilibrium mode" of the Optima XLA/I analytical ultracentrifuge user interface, or *ad hoc* at any time during the SV run [21, 22]: As long as at least a single scan is available at each speed step, the precise times τ where speed changes commenced can be reconstructed from the combination of elapsed time t and $\int \omega^2 dt$ entries of the files i and $i - 1$ with $\omega_i(t_i) > \omega_{i-1}(t_{i-1})$ as

$$
\begin{aligned}
\tau_i =& \omega_i^2 \left(t_i^{(\text{sed})} - t_i \right) - \omega_{i-1}^2 \left(t_{i-1}^{(\text{sed})} - t_{i-1} \right) \\
& - \omega_{i-1}^2 t_{a,i} + \omega_{i-1} \dot{\omega} t_{a,i}^2 + \frac{1}{3} \dot{\omega}^2 t_{a,i}^3 \left(\omega_{i-1}^2 - \omega_i^2 \right)^{-1},
\end{aligned}
\tag{2.35}
$$

with $t^{(\text{sed})}$ as defined in Eq. (2.34) and using the abbreviation $t_{a,i} = (\omega_i - \omega_{i-1})/\dot{\omega}$ for the time required to change the rotor speed at the (constant) rate $\dot{\omega}$ [22].

Due to dynamic rotor stretching, for the analysis of experimental data involving different rotor speeds in the same run, it is necessary to back-transform the radial profiles from the laboratory frame onto the same fixed-solution column radius reference frame [21, 22]. This can be accomplished with known stretching modulus of the rotor, and compensating for changes in the calculated local centrifugal fields by small corrections to the effective rotor speed (Eq. (5.3) in Part I, Section 5.2.3.2).

> SEDFIT offers several tools for working with time-varying centrifugal fields in SV.
>
> Rotor speed schedules are specified in SEDFIT in a three-column ASCII file `speedsteps.txt` specifying, for each rotor speed step, 1) the time the speed change was initiated (in sec); 2) the new target rotor speed (in rpm); and 3) the rotor acceleration or deceleration rate (in rpm/sec). This file must be present in the same folder as the scan data files.
>
> Additional utilities include the visualization of the rotor speed time course in the `Display` menu, and in the `Options ▷ Loading Options and Tools` menu functions to extract rotor speed profiles from experimental scan data via Eq. (2.35), conversion of `speedsteps.txt` files to `.equ` centrifuge GUI files and *vice versa*, as well as a radial transformation to compensate translation of the solution column from rotor stretching.

2.2.5 Dynamic Density Gradients

So far we have assumed that the solution density and viscosity are constant throughout the cell, so that the macromolecular sedimentation coefficient and diffusion coefficient is a constant. However, for experimental reasons discussed in Part I, Section 2.3.3, sometimes a co-solute is present at sufficiently high concentration to contribute significantly to the solution density and viscosity. In this case, concentration gradients arising from the sedimentation of the co-solvent create spatio-temporal variations in the solution density and viscosity. This is typically the case at molar concentrations of salts, high concentrations of glycerol, or in the presence of other excipients in pharmaceutical formulations. Typically, data acquisition modes can be found such that these co-solvents do not contribute to the signal, but their effect on the macromolecular sedimentation process can be very significant.

The description of such processes requires the consideration of local density $\rho_{xp}(r,t)$ and viscosity $\eta_{xp}(r,t)$, respectively, which relates to the local co-solvent concentration $\chi_3(r,t)$ with empirically measured relationships $\rho_{xp}(\chi_3)$ and $\eta_{xp}(\chi_3)$ [18]. Molecular constants are the sedimentation and diffusion coefficients at standard conditions, $s_{20,w}$ and $D_{20,w}$, respectively. Thus, with Eq. (1.22) we can describe the coupled co-solute and macromolecular sedimentation as

$$\frac{\partial \chi_3}{\partial t} = -\frac{1}{r}\frac{\partial}{\partial r}\left(\chi_3 s_3 \omega^2 r^2 - D_3 \frac{\partial \chi_3}{\partial r} r\right)$$

$$\frac{\partial \chi_a}{\partial t} = -\frac{1}{r}\frac{\partial}{\partial r}\left(\chi_a s_{20,w} \frac{\eta_0}{\eta_{xp}(\chi_3)} \frac{(1 - \bar{v}_{xp}\rho_{xp}(\chi_3))}{(1 - \bar{v}_0\rho_0)} \omega^2 r^2 - D_{20,w} \frac{\eta_0}{\eta_{xp}(\chi_3)} \frac{\partial \chi_a}{\partial r} r\right)$$

$$(2.36)$$

where s_3 and D_3 are the observed sedimentation and diffusion coefficient of the co-solute [18]. The reverse influence of macromolecular sedimentation on the co-solute can typically be neglected because of the much lower macromolecular concentrations. At low rotor speeds co-solute gradients tend to be very shallow, due to their small molar mass, small sedimentation coefficient, and high diffusion coefficient. At high rotor speeds, however, the co-solute sedimentation is often quite significant, resembling patterns as shown on p. 33 in Fig. 2.5A or B, with relative concentration differences from meniscus to bottom amounting to 10% or more. At high co-solute concentrations, this translates into significant local density and viscosity variations, and in this case the effect on macromolecular sedimentation due to local changes in solvent viscosity and buoyancy must be accounted for.

The evolving shapes of the macromolecular boundary depend on the relative sedimentation velocities of co-solute and macromolecule. Typical features have been described in [18] and are highlighted in Fig. 2.13: Due to the depletion of co-solute at radii below its hinge point, the macromolecular boundary will initially migrate faster (relative to a hypothetical stationary co-solute), and then experience continuous deceleration as it migrates into the buildup of increasing co-solute concentration. At high macromolecular s_{xp}-values, the macromolecular boundary may out-run and escape the co-solute depletion at small radii, but will always encounter the slow-down past the co-solute hinge-point. This causes a small maximum at the leading edge of the sedimentation boundary, and a radially increasing macro-molecule concentration replacing much of the solution plateau and counteracting radial dilution. At sufficiently high co-solute concentration the local solvent density may eventually grow above the point of neutral buoyancy for the macromolecule, causing the latter to float. This condition is achieved first at the bottom of the cell and with time migrates to smaller radii toward a point in the cell determined by the equilibrium co-solute distribution. With macromolecular sedimentation at

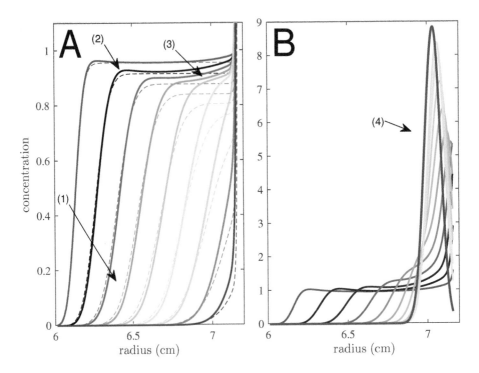

Figure 2.13 Examples for sedimentation boundary shapes in dynamic density gradient. *Panel A*: Simulated sedimentation profiles of a 117 kDa, 6.6 S-protein ($\bar{v}_{xp}$=0.73 ml/g, $s_{xp} = 4.3$ S) sedimenting at 50,000 rpm in the presence of a dynamic density gradient established from the sedimentation of 1.25 M CsCl. Protein sedimentation boundaries in time intervals of 1800 sec are shown as solid lines. For comparison, the hypothetical protein concentration profiles if the solution density and viscosity were constant at the initial loading conditions, i.e., neglecting co-solute sedimentation, are also shown as dashed lines. *Panel B*: Sedimentation of a large protein complex (500 kDa, 27 S, $\bar{v}_{xp}$=0.73 ml/g, $s_{xp} = 4.3$ S) in time intervals of 1800 sec, in the presence of a higher co-solute concentration (2.5 M CsCl) that develops solution densities after start of sedimentation that exceed the point of neutral buoyancy for the protein. Highlighted features are (1) the initially accelerated sedimentation, especially the trailing part of the boundary, due to co-solute depletion; (2) the transient maximum in the leading edge of the boundary from the concentration co-solute gradients; (3) lack of solution plateau due to continuously smaller macromolecular sedimentation at higher radii; (4) if neutral buoyancy densities are exceeded, flotation from the bottom of the solution column into a band.

small radii and flotation at high radii the result is the formation of a band that in equilibrium will have approximately Gaussian shape.

All of these features can be exploited to determine the macromolecular apparent partial-specific volume $\bar{v}_{SP}$ [18], and the analysis of the Gaussian allows the determination of the buoyant molar mass [101].[16] It is important to note that these quantities are significantly dependent on the co-solute identity and concentration,

---

[16]This approach has been developed by Meselson, Stahl, and Vinograd for the study of large DNA and viruses. Exploiting the high resolution of the method for measuring macromolecular density, it was applied by Meselson and Stahl in their seminal work on DNA replication [102].

due to hydration and preferential solvation effects (as discussed at length in Part I, Chapter 2).[17] The practical implementation of SV analyses considering dynamic density gradients from sedimenting co-solutes is discussed in Section 8.2.7 below.

In SEDFIT the sedimentation process with dynamic density gradients can be simulated and fitted to experimental data after invoking the Options ▷ Inhomogeneous Solvent model. It will cause macromolecular sedimentation parameters to be in units of standard conditions, which will be locally corrected to experimental conditions dependent on local density and viscosity.

2.3 THE TRANSPORT METHOD

2.3.1 Mass Balance in Sedimentation Velocity

The consideration of the mass balance in the total transport of particles provides a further important aspect of sedimentation, leading to an extremely powerful approach for the data interpretation. Basically, rather than trying to track the boundary position — which will be reported on by only the minority of particles in the cell — we want to know the rate of particles in the plateau region crossing an imaginary plane at a certain radius (Fig. 2.14) [103–106].[18] This approach is equivalent to counting the number of particles, or integrating the signal, above (i.e., at smaller radii than) a certain radius, and evaluating the change of this mass balance with time. This promises to utilize the signal from the majority of particles in the sample, and therefore provide excellent statistical precision. Additionally, as we will see, this approach can circumvent the need to solve the Lamm equation.

To develop the concept illustrated in Fig. 2.14 we can again use the relationship between the flux Eq. (2.6), defined as the number of particles per unit time crossing a unit area at radius r', with the concentration and velocity at that radius $j = c(r')v(r')$ as in Eq. (2.5) and arrive at

$$j = \frac{1}{A(r')} \frac{dN}{dt} = c_p s_j(r')\omega^2 r', \tag{2.37}$$

where $A(r')$ is the cross section of the solution column at radius r' in the plateau

---

[17]While in theory this would introduce a dependency $\bar{v}_{xp}(\chi_3(r,t))$ on the dynamically evolving local co-solute concentration, considering that the absolute co-solute concentration differences across the solution column are only on the order of $\sim$10 %, this is a second-order effect that may be neglected compared to the effect of co-solute concentration variations on solvent density.

[18]Historically, the basic concept was mentioned first by Tiselius, Pedersen and Svedberg [103] in conjunction with the presentation of a centerpiece containing a sieve aiding in preparative separation of solution contents after sedimentation. Gutfreund and Ogston [104] have further developed the analytical aspect of this approach, which were further expanded by Goldberg [105] and [106]. The approach was reviewed by Schachman [2] and Fujita [24], and revisited in modern context of direct boundary modeling in [107].

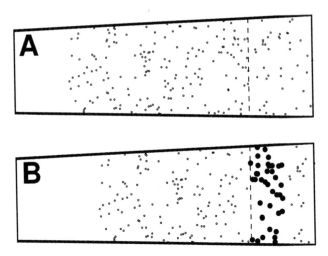

Figure 2.14 Basic concept of using mass balance to determine the sedimentation coefficient in the transport method. Shown are sedimenting point particles randomly distributed throughout the solution column at a certain point in time (here shown in normalized sedimentation time $\tau = 0.2$) (A) and then the same points propagated according to Eq. (2.2) to a later point in time $(\tau = 0.3)$ (B). In contrast to Fig. 2.1, where the sedimentation was followed by tracking the boundary position, in the transport method we assess how many molecules per unit time cross an imaginary line at a radius r' in the plateau region (dashed line). The particles crossing this line in the time interval between the top and bottom plot are indicated as filled circles.

region. Preliminarily we refer to the s-value so determined as $s_j(r')$ to indicate its origin from the local flux. On the other hand, due to mass conservation, the number of particles crossing the area must be identical to the change in the total number of particles remaining at radii above (i.e., at lower radii) than r'

$$\frac{dN}{dt} = -\frac{d}{dt} \int_{m}^{r'} \chi(r,t) h\phi r \, dr \,. \tag{2.38}$$

The integration limits are set to encompass the entire solution column from the meniscus (even though the solvent plateau, if formed, will not contribute) up to our reference radius r' in the solution plateau. Since the integration limits are constant in time, we can draw the time-derivative into the integral, and substitute $d\chi/dt$ with the Lamm equation Eq. (2.30), leading to

$$\frac{dN}{dt} = \int_{m}^{r'} \frac{1}{r}\frac{\partial}{\partial r}\left(\chi s\omega^2 r^2 - D\frac{\partial\chi}{\partial r}r\right) h\phi r \, dr \,, \tag{2.39}$$

which can be trivially integrated, resulting in the difference in total flux at the plateau radius r' and the meniscus. The latter disappears due to the impermeable boundary conditions at the meniscus, and with the choice of r' in the plateau, the diffusion term in the former disappears, too. We are left with[19]

$$\frac{dN}{dt} = c_p s \omega^2 (r')^2 h\phi \,. \tag{2.40}$$

From the comparison of this result with Eq. (2.37) and the area $A(r') = h\phi r'$ we find the important result

$$s_j(r') = s \,, \tag{2.41}$$

which establishes the link between the rate of depletion of material above a reference radius in the plateau and the molecular sedimentation coefficient, namely that the s-value determined from volume integration of the signal and considering mass balance is exactly identical to that of the sedimenting particles. Furthermore, it is obvious now that this value does not depend on the particular choice of our hypothetical radius r'.

2.3.2 The Weighted-Average Sedimentation Coefficient

The consequence of mass conservation can be rephrased and generalized: Combining Eqs. (2.37) and (2.38) we can write

$$\frac{1}{a'(r',t)\omega^2 r'^2}\left(-\frac{d}{dt}\int_m^{r'} a'(r,t)r\,dr\right) =: s_w^t \,, \tag{2.42}$$

defining an instantaneous weighted-average sedimentation coefficient [24,107]. Here, the concentration is substituted with the signal, which is equivalent in this equation because the multiplication with signal coefficient and optical pathlength cancels out. However, the signal is qualified as a' to highlight that time-dependent baseline offsets $\beta(t)$ are incompatible with Eq. (2.42), since they would create adventitious contributions to the temporal derivative of the integral. Even though the time-derivative is invariant with regard to radial-dependent baseline offsets $b(r)$, the reference to the plateau signal requires the absence of such offsets, too.

---

[19] Alternatively, for non-diffusing particles we can arrive at the same result by integration of Eq. (2.38) over the explicit concentration profile Eq. (2.10). Using the symbol $r^{(P)}(t)$ for the boundary position Eq. (2.2) and $c_p(t)$ following Eq. (2.8) for the solution plateau:

$$\begin{aligned}
\frac{dN}{dt} &= -h\phi\frac{d}{dt}\left(c_p\int_{r^{(P)}}^{r'} r\,dr\right) \\
&= -h\phi\left(-2\omega^2 s c_p\left(\frac{1}{2}(r')^2 - \frac{1}{2}(r^{(P)})^2\right) + c_p\frac{dr^{(P)}}{dt}\frac{d}{dr^{(P)}}\int_{r^{(P)}}^{r'} r\,dr\right) \\
&= -h\phi\left(-2\omega^2 s c_p\left(\frac{1}{2}(r')^2 - \frac{1}{2}(r^{(P)})^2\right) + c_p s\omega^2 r^{(P)}\times\left(-r^{(P)}\right)\right) \\
&= h\phi c_p\omega^2 s(r')^2 \,,
\end{aligned}$$

which is identical to Eq. (2.40), consistent with expectation.

With this caveat, Eq. (2.42) can be directly applied to experimental data solely by considering the relative change in the volume integral over the observed boundaries: *It does not require a particular physical motivation or mathematical model to fit the data — any model that faithfully describes the volume integrals under the sedimentation boundaries can provide the correct weighted-average s-value.*

Further, it is worth noting that in this definition (as in the derivation above) the use of the Lamm equation solely relied on the instantaneous local fluxes, but neither the initial conditions of sedimentation were referenced, nor the absolute time since the start of the sedimentation.[20] This absence of history-dependence provides opportunities to extract sedimentation coefficients from experiments where adventitious experimental conditions, such as from initial convection, were present but have subsided prior to the mass balance analysis.

The potential time dependence of s_w^t is inherited from the time dependence of the sedimentation boundaries. We recall that the derivation of the identity Eq. (2.41) was based on equations for a single sedimenting species. If this indeed reflects the sedimentation experiment producing the boundaries $a(r,t)$ in Eq. (2.42), then $s_w^t = s$ at all times, but when Eq. (2.42) is used as an operational definition to characterize other sedimentation processes, more complex situations may arise.

SEDFIT offers two variations of calculating s_w^t corresponding to Eq. (2.42) in the Options ▷ Calculator menu: ▷calculate weight-average s by second moment integration and ▷calculate weight-average s by differential second moment integration. The former uses the transport method to determine s from mass balance relative to the initial loading concentration, whereas the latter is not history dependent and solely evaluates the instantaneous rate of change. Both require all systematic noise to be removed. The radius r' is specified by the right fitting limit, with the assumption that the plateau extends at least 1 mm toward smaller radii from r'.

We refer to s_w^t as a "weighted average" as it can be easily seen that for mixtures of particles with different s-value, s_w^t will be an average between those of the sedimenting species, since all derivations above are linear in particle concentration. For example, for mixtures of two kinds of particles in Eq. (2.40) the total mass balance is $dN^{(\text{tot})}/dt = -h\phi\omega^2 r'^2(s_1 c_{p,1} + s_2 c_{p,2})$. Since the total plateau concentration is $c_{p,tot} = c_{p,1} + c_{p,2}$, after consideration of the signal coefficient ε_λ, the definition Eq. (2.42) will result in

$$s_{w,\lambda}^t = \frac{s_1 \varepsilon_{1,\lambda} c_{p,1}(t) + s_2 \varepsilon_{2,\lambda} c_{p,2}(t)}{\varepsilon_{1,\lambda} c_{p,1}(t) + \varepsilon_{2,\lambda} c_{p,2}(t)} \tag{2.43}$$

(with the optical pathlength canceling out, but not signal coefficients) and more

---

[20]However, configurations are excluded from the definition Eq. (2.42) that have concentration gradients at r' (since diffusion fluxes would not vanish), and trivially those that lead to vanishing concentrations at the radius r'.

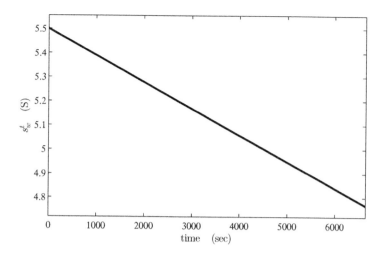

Figure 2.15 The time dependence of the s_w^t-value defined by the transport method via the total flux in the plateau region. As an example, s_w^t was calculated for two species of 1 S and 10 S, respectively, both at unit loading concentration and with a unit signal increment. Because the radial dilution proceeds faster for the 10 S species, the relative contribution of this species to the signal in the plateau diminishes with time. Therefore, instead of observing the correct average s-value of 5.5 S, we find s_w^t decreasing strongly with time. The time span shown ranges from the beginning of sedimentation in a 12-mm solution column until the faster boundary has reached the bottom of the solution column at 7.2 cm.

generally

$$s_{w,\lambda}^t = \frac{\sum_i s_i \varepsilon_{i,\lambda} c_{p,i}(t)}{\sum_i \varepsilon_{i,\lambda} c_{p,i}(t)} , \tag{2.44}$$

i.e., we obtain the signal-weighted average sedimentation coefficient of all particles i in solution.

It is generally not well appreciated that for heterogeneous systems the s_w^t-value based the definition Eq. (2.42) will depend on the time point during the experiment at which the transport is considered.[21] In fact, s_w^t is usually identified directly with s_w, the weighted average s-value. However, because the plateau concentrations of each species decrease exponentially depending on their particular s-value, following Eq. (2.8) and the square dilution law Eq. (2.9), the relative contribution of the faster sedimenting species will decrease at later times relative to the slower sedimenting species. An example of the magnitude of this decrease for two sedimenting species of 1 S and 10 S at equal loading signals is shown in Fig. 2.15: Instead of the constant average of 5.5 S, the s_w^t-value drops from 5.5 S initially to 4.77 S at the time when the plateau of the faster species would disappear. This change by ∼13% is far above the measurement error and would make s_w^t a poorly defined and ambiguous quantity when studying mixtures.

Therefore, in order to arrive at value for s_w that unambiguously characterizes

---

[21]This effect is different from the time dependence of s_w in interacting systems due to radial dilution causing chemical re-equilibration, as discussed in [3, 107].

the sample as it was initially loaded, unaffected by the differential size-dependent dilution during the SV experiment, we amend the definition of s_w^t in Eq. (2.42) and take the limiting value at the start of centrifugation:

$$s_w =: \lim_{t \to 0} \left[\frac{1}{a(r', t)\omega^2 r'^2} \left(-\frac{d}{dt} \int_m^{r'} a'(r, t) r \, dr \right) \right]. \qquad (2.45)$$

We refer to this as the weighted-average sedimentation coefficient s_w. Likewise, it follows

$$s_{w,\lambda} = \lim_{t \to 0} \left[\frac{\sum_i s_i \varepsilon_{i,\lambda} c_{p,i}(t)}{\sum_i \varepsilon_{i,\lambda} c_{p,i}(t)} \right]. \qquad (2.46)$$

Defined like this, s_w is a quantity that is invariant on experimental or data analysis conditions, and describes a property of the sample only. For interacting systems, it can be interpreted in a thermodynamic context of composition-dependent populations of species generating isotherms of s_w.

In the literature, s_w is often referred to as a "weight-average" s-value, assuming that the molar signal increments of all species are proportional to their molar mass.[22] However, this is rarely strictly the case, in spite of this being a common approximation for interference optical detection of proteins (corresponding to the assumption that the weight-based refractive index increment of all proteins is the same [108]). Furthermore, this assumption can be far from accurate for absorption optical detection, in particular when studying heterogeneous mixtures. Therefore the term "weighted" average (or sometimes "signal-average") seems more precise, reminding us of the need to always consider the particular signal increments of all species in optical detection system used. Thus, the same sample may exhibit different s_w-values at different wavelengths, and this can provide valuable information on sample composition.

In summary, the transport method for determining s_w is a very robust strategy to extract precise information from the sedimentation process, without explicit reference to the shape of the sedimentation boundary. An important consequence is that any boundary model that accurately reflects the integral under the radial profiles will conserve the information on s_w and permit its accurate determination. The signal profiles are rarely directly integrated (see below), because of the difficulty of assigning the correct baseline parameters (RI and TI noise). Instead, the integrals over the signal profiles are better evaluated in a secondary step following the direct fit of the raw data with a boundary model. *Any model that fits the data well will preserve the integral and therefore allow the precise determination of s_w, irrespective of the nature of the model, and whether it is empirical or physically accurate.*

---

[22]A historic root of this is the study of polymers where different averages, including number average, z-average, and weight average, are used to characterize the polydisperse distributions. The continued utility of these concepts when studying mixtures of intrinsically discrete biological macromolecules is unclear, in light of modern analysis techniques that are able to characterize explicitly entire distributions with relatively high hydrodynamic resolution.

In particular, we will show in Section 3.4 how the s_w-value can be extracted from continuous sedimentation concentration distributions derived from direct boundary fitting, such as the $c(s)$ and $ls\text{-}g^*(s)$ family. This provides the highest precision.

2.3.3 Second Moment Analysis

For diffusionally broadened boundaries, one can ask if there is a method to determine the precise points $\bar{r}$ for each boundary that would correspond to an infinitely sharp boundary moving with the same s-value (or s_w^t, to be more accurate). Goldberg has shown that such a point exists, and that it is the second moment of the gradient curve [105]

$$\bar{r} = \frac{\int_m^{r'} r^2 (d\chi/dr) dr}{\int_m^{r'} (d\chi/dr) dr} \tag{2.47}$$

from which follows by partial integration

$$\bar{r}(t) = \sqrt{r'^2 - \frac{2}{c_p} \int_m^{r'} \chi(r,t) r \, dr} \tag{2.48}$$

(for boundaries that have separated from the meniscus with $\chi(m) = 0$). Similar to s_w in the transport method, the result is independent of r' as long as r' is in the plateau region. $\bar{r}(t)$ follows the relationship for the propagation of non-diffusing boundaries Eq. (2.3) with sedimentation coefficient s_w^t,[23] and, accordingly, a plot of $log(\bar{r}/m)$ vs. t will have a slope $\omega^2 s_w^t$. For heterogeneous systems, however, the time dependence of s_w^t represents a problem in theory leading to a curvature of $log(\bar{r}/m)$ vs. t.

For many reasons, this seems obsolete as a practical strategy to obtain s_w-values from experimental data. However, its appeal is that it provides a rigorous avenue for using the simple expression for the migration of non-diffusing boundaries to determine s (via a plot of $log(\bar{r}/m)$ vs. t), at least for homogeneous single-species boundaries, despite the presence of diffusion in the experimental boundaries, and without need to interpret the boundary shape.

---

[23]We have

$$\frac{d}{dt} log\left(\frac{\bar{r}}{m}\right) = \frac{d}{dt} log(\bar{r}) = \frac{1}{\bar{r}} \frac{d\bar{r}}{dt} = \frac{-1}{2\bar{r}^2} \frac{d}{dt}\left[\frac{2}{c_p}\int_m^{r'} \chi(r,t) r \, dr\right]$$

$$= \frac{-1}{2\bar{r}^2}\left(\left[\frac{d}{dt}\frac{2}{c_p}\right]\int_m^{r'} \chi(r,t) r \, dr + \frac{2}{c_p}\left[\frac{d}{dt}\int_m^{r'} \chi(r,t) r \, dr\right]\right).$$

With Eqs. (2.8) and (2.48) simplifying the first term, and the definition of s_w^t Eq. (2.42) for the second term, this simplifies to

$$\frac{d}{dt} log\left(\frac{\bar{r}}{m}\right) = \frac{-1}{2\bar{r}^2}\left((-2\omega^2 s_w^t)\left(\bar{r}^2 - r'^2\right) - 2\omega^2 s_w^t r'^2\right)$$

$$= \omega^2 s_w^t$$

as in Eq. (2.3).

In practice, different approaches have often been used to determine $\bar{r}$ avoiding the integration Eq. (2.48). It should be noted that both the point at the half-maximum signal, as well as the point of steepest slope would be an incorrect choice [86].[24] However, just like any fitted model that preserves the volume integral will lead to the correct s_w *via* the transport method, any empirical fitting model to sedimentation boundaries that preserves the second moment boundary position can be used to determine $\bar{r}$. As shown by Attri, Lewis and Korn [109] an example for such an empirical function preserving $\bar{r}$ is

$$a(r) \approx \frac{\exp\{k(r^2 - \bar{r}^2)\}}{1 + \exp\{k(r^2 - \bar{r}^2)\}} + b \qquad (2.49)$$

with k denoting an empirical boundary width parameter.[25]

---

[24]This initially may seem counter-intuitive, but arises from the sector-shaped geometry of the sedimentation process. Errors of taking, for example, the points of maximum slope are small for single, ideally sedimenting species, but may be much larger when the boundary exhibits unresolved heterogeneity. In this case, the resulting s-value can be a poorly defined intermediate of the s-values of the species present in the boundary.

[25]For example, superpositions of the r.h.s. of Eq. (2.49) were used for multiple boundaries, each followed by the determination of s_w-values by $log(\bar{r})$ *vs.* t, in the analysis of sedimentation boundaries of actin oligomers [109].

CHAPTER 3

Properties of Sedimentation Coefficient Distributions

D ISTRIBUTIONS are essential tools for the description of most SV experiments due to the immense information content and exquisite sensitivity of SV data for sample heterogeneity. For example, it is quite rare that a biological sample, such as a protein preparation, is truly monodisperse, free of breakdown products, and does not contain at least low levels of aggregates detectable by SV. Further, many of the more interesting experimental questions are directed at the formation of new species due to attractive interactions between different components in a mixture. In other disciplines, such as polymer chemistry, the need for models that capture the polydisperse nature of the particle ensemble under study is even more obvious, due to the lack of a highly specific synthesis machinery producing monodisperse macromolecules.

The treatment in the present chapter assumes that there is no interaction between the different particles in the mixture. We assume here that particles are sedimenting independently and ideally, in the limit of infinite dilution, i.e., without concentration-dependent repulsive interactions. In this case, the total signal is a sum of Lamm equation solutions of the form described above, for a range of species potentially in solution. The problem is to unravel the individual signal contributions of each species given the noisy total signal.

The rigorous distribution analysis rests on mathematical principles that are explicitly stated and can be intuitively rationalized. Most computational details are deferred to Appendix B. They consist of well-established standard mathematical operations that can be executed with common personal computers. Limitations and approximations are explicitly discussed.

3.1 OUTLINE OF THE PROBLEM

It is invaluable to develop some physical intuition for the relationship between sedimentation coefficient distributions and the visual appearance of sedimentation patterns. This can give confidence in the results from the mathematical analyses, and allows anticipating the presence of parameter correlations and alternative interpretations where the families of observed sedimentation profiles between different types of distributions are similar.

The presence of multiple species will lead to superpositions of sedimentation boundaries discussed in the previous chapter. The key question for the data analysis is whether we can diagnose their presence and resolve the individual contributions from noisy data. The resolution of boundaries from different species is a strong function of the rotor speed: increasing the rotor speed decreases the time required for the particles to traverse the sample cell and therefore diminishes the diffusional transport. Unfortunately the range of experimentally available rotor speeds is restricted for safety reasons; in the current commercial instruments the maximum rotor speed rating of the different rotors is 50,000 rpm for the 8-hole rotor and 60,000 rpm for the 4-hole rotor.[1]

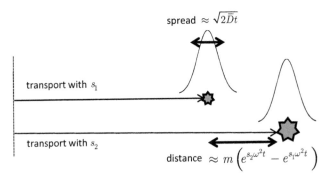

Figure 3.1 Schematics of the movement of two species of particles with different sedimentation coefficients. After a certain time t, sedimentation has transported the boundary midpoints of both species from the meniscus (vertical dotted line on the left) to different positions. Their separation has grown to $m(e^{s_2\omega^2 t} - e^{s_1\omega^2 t})$. Each boundary is blurred from diffusional broadening, which during the same time has reached an r.m.s. width in the order of $\sim \sqrt{2\bar{D}t}$. If the separation between the boundary midpoints is larger than this amount, then one can observe without difficulty two separate boundaries. If, on the other hand, the diffusional spread is larger than the boundary separation, then the two boundaries merge into one broad boundary. The spread of the combined boundary is much faster than $\sqrt{Dt}$ and follows a different evolution with time.

The basic problem has been at the center of SV analysis for many decades [110, 111]: Let us consider only two different classes of particles with sedimentation coefficients s_1 and s_2. The separation of the boundary midpoints is $m(e^{s_2\omega^2 t} - e^{s_1\omega^2 t})$, and the r.m.s. displacement from diffusion is approximately $\sqrt{2\bar{D}t}$ (with $\bar{D}$ denoting an average diffusion coefficient)(Fig. 3.1). If the separation between the boundary midpoints of both species is larger than the blurring of each boundary from diffusion, then we expect two visually distinct boundaries to appear.

---

[1]In our laboratory we did not find it necessary to limit the rotor speed to the rating of the centerpieces. In our experience, the only downside of using charcoal-filled Epon centerpieces at 50,000 rpm or 60,000 rpm is that one should be confident that there is no leak, which at high speeds could cause breakage of the septum dividing the sectors (see Part I, Section 5.2.4.1).

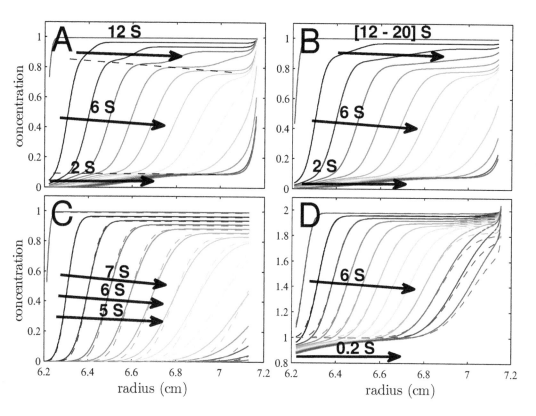

Figure 3.2 Examples for the sedimentation profiles of mixtures of non-interacting particles. For all simulations, the main boundary is formed by a 100 kDa, 6 S species with different polydispersity. *Panel A*: With distinct impurities from a 20 kDa, 2 S species and a 250 kDa, 12 S species, all boundaries are well resolved, as indicated by the dashed lines highlighting the transition region between the different boundary components. *Panel B*: Using the same conditions as *A*, but substituting the discrete 12 S species with a broad, uniform distribution of particle sizes with *s*-values between 10 S and 20 S causes sloping of the apparent plateau region of the main species. *Panel C*: The same 100 kDa, 6 S species in superposition with an 80 kDa, 5 S species and a 120 kDa, 7 S species, both at 20% of the total signal allows no clearly separate boundary components to be visually discerned. However, the boundary is significantly broader than would be obtained with a pure 100 kDa, 6 S species (for comparison shown as dashed lines). *Panel D*: The 100 kDa, 6 S species is shown in superposition with an equal signal arising from the slow sedimentation of a 0.5 kDa, 0.2 S species. The latter has a very high diffusion coefficient, causing a rapidly evolving deflection of the signal in both the meniscus and the bottom region (compare Fig. 2.5A), which increases the apparent width of the macromolecular boundary component. For visual comparison, the contributions from the macromolecular component alone are shown as dashed lines. For all simulations, the sedimentation profiles are calculated under conditions of phosphate buffered saline at 20°C, for a rotor speed of 50,000 rpm, with the first scan shown at 300 sec and all following in 900-sec intervals (*A–C*) or for 40,000 rpm in constant 1000 sec intervals (*D*).

Otherwise, only one combined boundary will be formed, which appears much broader and evolves much faster with time than predicted by diffusion alone.

Fig. 3.2 shows different examples for these two cases. If the sedimentation rates of each species are very different such that the separation of their boundaries is much larger than their diffusional spread, multiple boundaries can be readily visually recognized. In the illustration of this case in Fig. 3.2 Panel A, in addition to the main 6 S species a slower and a faster sedimenting species is clearly present. Trivially, the total signal is the superposition of the sedimentation boundaries of each species, in a way that does not lead to distortion or likely misinterpretation of any of the individual boundaries.[2]

The second case, where the difference between the s-values of the particles in the mixture is smaller such that they do not separate into distinct boundaries, is shown in different variations in Fig. 3.2 Panels B–D. It is characterized by mixed boundaries that more rapidly broaden with time.

Panel B illustrates the family of concentration profiles that is typically observed when there is not a single large species, but a broad distribution of aggregates, which individually cannot be resolved. This case can be qualitatively reminiscent of a sloping plateau from the main boundary component, for which there can be other causes.[3] Whether the larger species can be individually resolved or not, it is clear that the magnitude of any deviation from the expected constant plateau value will be detectable with high precision, considering that the experimental data provide many points with high signal/noise ratio in the plateau region. This is the reason for the high sensitivity of SV for the detection of trace aggregates.

A different example is presented in Fig. 3.2 Panel C, which shows a main species of 100 kDa and 6 S, in addition to 20% of the total signal from an 80 kDa species with 5 S, and 20% of the total signal originating from a 120 kDa species with 7 S. It is instructive to compare the resulting total signal to that of a pure 100 kDa, 6S species (dashed line in Fig.3.2C), which has much steeper boundaries. Clearly, if heterogeneity would not be recognized as a major contributor to the boundary spread, and the data incorrectly modeled as if there was only a single diffusing species, then the excess boundary spread would lead to a large overestimate of the diffusion coefficient, D^*. If s and the apparent D^* are inserted into the Svedberg equation Eq. (1.35) to estimate the molar mass, the calculated M^*-value will be a gross underestimate. For the present case, the result would be ~65 kDa, rather than the true average of 100 kDa (see Fig. 3.3 below). Therefore, *in the absence of hydrodynamic non-ideality, single-species models in SV without resolving the boundary heterogeneity will result only in lower bounds for the true molar mass.*

---

[2] As is apparent from Fig. 3.2 Panel A, the lower the s-value of the smallest species, the longer the time of data acquisition should be. A full characterization would require data from about twice as long as the time range shown, such that the smallest species can migrate across the majority of the cell.

[3] These may include, for example, signals from excess buffer salts in the sample or reference sector in conjunction with interference optical detection; radially increasing buoyancy due to solvent compressibility or dynamic density gradients of small co-solutes; mis-alignment of the monochromator leading to a radial-dependent detection wavelength in the absorbance optics; and radial signal magnification gradients in the fluorescence optics.

This is another reflection of the exquisite sensitivity for heterogeneity, which may be exploited for judging the purity of a preparation.

Finally, Panel D of Fig. 3.2 provides an example when merging boundaries occur not due to closely spaced s-values, but due to very large diffusion coefficients. For species smaller than 1 kDa the migration driven by diffusion is much faster than that of sedimentation; for example, it can be of similar magnitude as the sedimentation transport of a 100 kDa protein. Such superposition of signals from small and large species may also occur, for example, when unmatched buffer salts contribute to the signal in the interference optical detection, or when small species contribute to the absorbance detection such as nucleotides in UV, or as a result of unbound chromophoric labels detected in the VIS (see Part I, Section 4.3.4). It is noteworthy that in such cases deflections and boundary broadening occurs, in particular, in the early scans and late scans close to meniscus and bottom, respectively. If unrecognized, this will lead to errors in the sedimentation parameters of the main species.

It is obvious that the problems outlined so far are exacerbated when even more components are present in solution, or for truly continuous distribution of sizes. Of the different types of sedimentation behavior of mixtures listed above, the data in

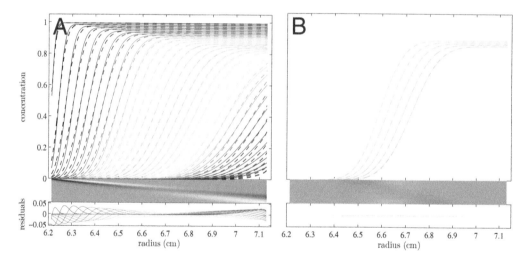

Figure 3.3 Distinguishing boundary broadening processes from diffusion and heterogeneity. As a model data set we consider the unresolved three-species mixture of Fig. 3.2C (a 100 kDa, 6 S species in superposition' with an 80 kDa, 5 S species and a 120 kDa, 7 S species, both at 20% of the total signal), and carry out a fit with an impostor single-species model. *Panel A*: Using data representative of the entire sedimentation process, the impostor model leads to a poor fit with systematic residuals of rmsd 0.0085 ranging from −0.055 to 0.04. The best-fit molar mass is 65 kDa, significantly below the main 100 kDa species, and significantly lower than the average of all species. *Panel B*: When using data from only a small time interval (compare Section 4.4), a relatively good fit is obtained with residuals with an rmsd of 0.0018, ranging between −0.0055 and 0.004, which would be below the typical signal/noise ratio of data acquisition. The best-fit molar mass is 66 kDa, still far below the average value of all species of 100 kDa.

Fig. 3.2C with species of rather similar s-values are among the most difficult cases, since the heterogeneity is not obvious to the experimenter's eye. If the presence of multiple species is not recognized, the exact amount of the underestimate of the best-fit molar mass M^* depends not only on the number, s-values, and populations of the contaminating species, but also on the particular data range chosen for the incorrect model. Importantly, the degree of misfit from using the incorrect model depends very much on the range of data selected in the analysis: the smaller the data set — in particular with regard to its time interval from the first to last scan considered — the better the impostor fit, and the less likely the error can be recognized.

This is illustrated in Fig. 3.3B, where a small central data subset provides a fit with deviations within the typical signal/noise ratio of the experiments. In some cases it may be possible to flag the incorrect model simply from the combination of M^* and s implying physically impossible frictional ratios f/f_0. However, in more subtle cases, this criterion will fail to help us recognize sample heterogeneity contributions to the boundary spread. On the other hand, if the complete time course of sedimentation is considered — from the first deflection at the meniscus until virtually complete depletion of material from the radial window that can be imaged, and if a large solution column is used — it becomes apparent that the evolution of the concentration profiles of Fig. 3.2C cannot be described well with a single-species diffusion model. As shown in Fig. 3.3A, in this case the amplitude of the residuals is ~10-fold above the usual signal/noise ratio. This highlights the importance of obtaining a good fit of the model to the entire experimental sedimentation boundary data. The inspection of the quality of the residuals is imperative for a reliable analysis. In the case of Fig. 3.3A, they would lead to a clear rejection of the impostor single-species diffusion model.

The underlying physical reason why it is possible to identify heterogeneity in the sedimentation boundary if it is observed over a large time interval is that migration from sedimentation proceeds exponentially with time $\sim e^{s\omega^2 t}$ whereas diffusion proceeds with a $\sim \sqrt{t}$ time dependence — the former accelerating transport with time and the latter resulting in a diminished net transport with time [112]. As a consequence, it is prudent to consider as large a time interval as possible, from the very beginning of sedimentation until all boundary components have migrated out of the observation window. This will be assumed in the following unless noted otherwise.

The examples of Figs. 3.2C/3.3 seem to be very drastic; however, it is not at all unrealistic or uncommon in its extent.[4] In fact, the detection and resolution of sample heterogeneity contributing to the spread of the total SV boundaries is a central problem in the analysis of SV. This was already recognized by Svedberg [113], and several approaches to address this issue have been reported by different investigators

---

[4]For example, a case study was presented [67] where a dimeric receptor molecule with some impurities and microheterogeneity due to glycosylation can appear as a monomer in an impostor single-species fit.

over many decades [14,57,112,114–120]. The modern strategy of distribution analysis outlined in the remainder of the chapter is based on having Lamm equation solutions readily available, in combination with current tools for the inversion of integral equations.

3.2 BASIC FRAMEWORK

3.2.1 Definition

Faithful to the principle of building a mathematical model for the observed data that is as physically accurate as possible and can be modeled directly to the experimental data, distributions of sedimenting particles are naturally described as an integral equation: Let us denote the family of experimental signal profiles as $a(r,t)$, and let $\chi_1(s,r,t)$ be the signal that would arise from a single species at a loading concentration of 1.0 (in the unit of the optical detection system, i.e., 1 fringe or 1 OD) sedimenting with sedimentation coefficient s. We can then write the definition of the simplest, generic sedimentation coefficient distribution, which we may call $\gamma(s)$, as integral equation

$$a(r,t) \cong \int_{s_{\min}}^{s_{\max}} \gamma(s)\chi_1(s,r,t)ds + b(r) + \beta(t) \tag{3.1}$$

(with the symbol $\cong$ denoting minimal squared residuals, and $b(r)$ and $\beta(t)$ denoting the baseline terms as outlined in Section 1.1.3.2). Eq. (3.1) means that the total signal is simply the sum of all signals arising from all subpopulations of particles with different sedimentation coefficients. The distribution $\gamma(s)$ is a differential sedimentation coefficient distribution, i.e., the product $\gamma(s)ds$ reports the population of particles with s-values between s and $s + ds$ in units of their loading signal at the beginning of the sedimentation process. Analogously, integration of the distribution between values s_1 and s_2 gives the loading signal of particles in that interval of s-values, and integration over the complete distribution

$$\int_{s_{\min}}^{s_{\max}} \gamma(s)ds = c_{0,\text{tot}} \tag{3.2}$$

gives the total loading concentration in signal units.

The distribution $\gamma(s)$ is generic in a sense that more specific details are necessary to derive a particular implementation of a sedimentation coefficient distribution. For example, since the Lamm equation depends for each species on two parameters, s and D, the concept above needs to be extended to two-dimensional distributions (such as the size-and-shape distribution $c(s, f/f_0)$ and its collapsed relative, the general $c(s,*)$). Alternatively, a rule for a functional dependence of $D(s)$ can be specified in a way that is appropriate for the particular mixture under study (such as in $c(s)$), or the approximation $D = 0$ may be used (such as in ls-$g^*(s)$). Further, to model data acquired in the presence of density gradients one could envision a distribution that includes particles of different partial-specific volume (e.g., $c(s, f/f_0, \bar{v})$). Such multi-dimensional distributions can be generally expressed as

$$a(r,t) \cong \int\limits_{p_{\min}}^{p_{\max}} \int\limits_{s_{\min}}^{s_{\max}} \gamma(s,p)\chi_1(s,p,r,t)dsdp + b(r) + \beta(t) \tag{3.3}$$

for two parameters s and p, and higher-order distributions can be envisioned analogously if p is a multi-dimensional vector. Examples for particular forms of such differential sedimentation coefficient distributions and their specific properties will be described below. However, all of them have several general properties in common that can be derived from the similar structure of these models. A crucial aspect of distribution models is that they are always actively constructed by the investigator, to match the needs for the experimental data and sample under study.

3.2.2 Range of the Distribution

The s-range of the distribution is given by the minimum and maximum values, $s_{\min}$ and $s_{\max}$, respectively. These values may be set to the widest theoretically possible range, but it is preferable in practice that they are empirically adjusted for the data under consideration.

In the absence of flotation, the smallest possible s-value is $s_{\min} = 0$, in which case $\gamma(s_{\min}, r, t)$ will become a constant baseline offset. It will be completely correlated with $b(r)$ and/or $\beta(t)$, respectively. $s_{\min}$-values slightly larger than zero, such as 0.1 S, are strictly not identical to a baseline offset (compare Fig. 2.5A), but may nevertheless exhibit a strong correlation with the baseline parameters. This problem is exacerbated in data sets comprising only short total experimental times. However, such a correlation may not be of further consequence as long as the value of $\gamma(s_{\min}, r, t)$ is not subjected to quantitative interpretation. The correlation with a baseline parameter can be suppressed using the Bayesian approach described below.

The largest theoretically sensible s-value can be calculated *via* Eq. (2.3) from the time point of the first available scan, $t_{\min}$, and the maximum radius in the observation window, $r_{\max}$, as

$$s_{\max} = \frac{1}{\omega^2 t_{\min}} log\left(\frac{r_{\max}}{m}\right). \tag{3.4}$$

Particles sedimenting faster could not possibly be observed as they would have migrated too close to the bottom of the cell already at the time of the first scan. In practice, $s_{\min}$ and $s_{\max}$ are usually chosen to confine the distribution to a narrower range, to just cover the range of s-values occurring in the sample, based on the available data. For detailed practical considerations on the choice of $s_{\min}$ and $s_{\max}$, see Sections 8.2.1 and 8.2.2, respectively.

3.2.3 Discretization

For the computational determination of distributions Eqs. (3.1) or (3.3), the problem needs to be expressed on a grid of discrete s-values, and take into account the discrete nature of the data points. This leads to the least-squares expression of the type

$$\underset{\gamma(s)\geq 0}{Min} \sum_{i,j} \left[a(r_i, t_j) - \sum_{l=1}^{N} \gamma(s_l)\chi_1(s_l, r_i, t_j)\Delta s_l - b(r_i) - \beta(t_j) \right]^2 , \qquad (3.5)$$

where indices i and j refer to a certain radial point and scan time, respectively, and the distribution is discretized in N values $\gamma(s_l)$ at s-values s_l from $s_1 = s_{min}$ to $s_N = s_{max}$. Uniform grid spacing $\Delta s_l = (s_{l+1} - s_{l-1})/2$ is usually chosen.[5,6] The choice of N should be guided by the quality of fit it can provide. Not shown in Eq. (3.5) are the regularization terms that need to be included in practice, as discussed below. Regularization makes the result largely independent of the choice of the resolution N, as long as N is sufficiently large so that the s-grid can smoothly describe the features of the distribution and accomplishes a fit within the noise of data acquisition.

3.2.4 Fitting Combinations of Model Functions to Experimental Data

The general structure of the problem of determining a sedimentation coefficient distribution is that of inverting a Fredholm integral equation of the first kind. In integral equations, the signal in the data space that would originate from a monodisperse distribution (i.e., from a δ-function) is termed the "kernel." The kernel in the present case is the Lamm equation solution. The integral equation simply corresponds to the question: What is the combination of amplitudes of theoretical Lamm equation solutions at different s-values that would fit the data best? Fig. 3.4 illustrates this, in a way making it obvious which s-values would need to be combined, and in which proportions. The combination of best-fit amplitudes then constitutes the sedimentation coefficient distribution. Intuitively, one could perhaps

---

[5]Note that the grid of discrete s-values does not have to be equidistant. In fact, it can be freely specified. When a large range of s-values over several decades should be spanned, it can be advantageous to apply a logarithmic or custom grid.

[6]In practice, it can occasionally be convenient to leave out the multiplication with the grid interval Δs. This will cause the distribution not to be normalized, i.e., to represent directly the best-fit amplitudes of the kernel (i.e., the Lamm equation solution) at each grid point. The amplitudes of non-normalized distributions will depend on the particular grid spacing. However, if an equidistant grid is used, the resulting distribution will be proportional to the standard normalized distribution. Non-normalized distributions relate to the concentration by summation over the distribution values at the grid points, rather than integration. Non-normalized distribution can be useful, for example, in the context of hybrid discrete/continuous distributions for better visual comparison of the concentration of the discrete species with the signal in the continuous sections, or for size-and-shape distributions eliminating (or deferring) the scale factors in the transformation between different coordinate systems (see Section 5.1).

compare the inversion of integral equations with the problem of the decomposition of an experimental spectrum into an unknown combination of known basis spectra (with the theoretical Lamm equation solutions at different s-values in this picture corresponding to the signal patterns of the basis spectra).

Even though the details can be deferred to Appendix B for the interested reader, it is very useful to be familiar with the basic mathematical principle of how the best combination is found, and that it rests on unambiguous solutions derived from standard calculus and matrix algebra: After discretization of the problem, the minimization in Eq. (3.5) can be solved by taking the partial derivative with respect to any particular concentration $\gamma(s_k)$, which — just as in one-dimensional calculus — must vanish at the minimum (Eq. (B.2)). Basic calculus then leads to a

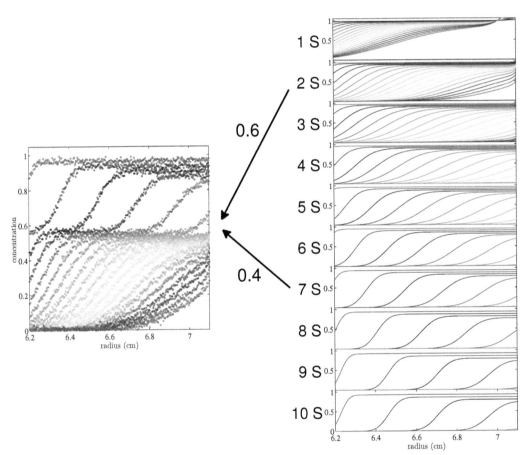

Figure 3.4 Concept of sedimentation coefficient distribution analysis. The noisy experimental data (left) are compared with a series of theoretical Lamm equation solutions at different s-values (right). To highlight the principle, each family of concentration profiles is taken at the same time points and displayed with the same color scheme. The least-squares fit determines the combination of amplitudes that fit the experimental data best. This is here $\gamma(2) = 0.6$ and $\gamma(7) = 0.4$, as indicated by the arrows, while otherwise $\gamma(s) = 0$. This set of amplitudes constitutes the sedimentation coefficient distribution.

simultaneous equation system for all $\gamma(s_k)$ with coefficients arising from summation over products of data points and kernels (Eq. (B.5)), which can be solved with standard numerical methods.

Perhaps the best illustration for the relationship between the experimental data and the amplitudes of the sedimentation pattern for the different s-values is the ls-$g^*(s)$ distribution of non-diffusing particles [121]. In this special case, the Lamm equation solutions kernels degenerate to step functions (Fig. 2.3), which are combined to form stair case-shaped concentration profiles as models of the experimental data. The height of each step directly displays the best-fit amplitudes of each Lamm equation solution with different s-value, as shown in Fig. 3.5. The possibility of easily recognizing this relationship visually usually vanishes when diffusional contributions for each species are accounted for, and the steps merge.

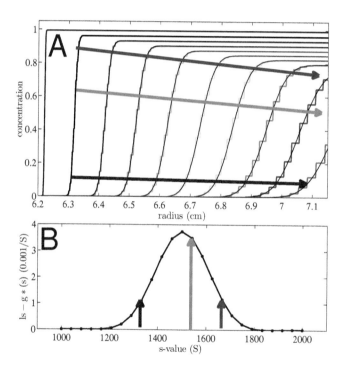

Figure 3.5 Illustration of the concept of ls-$g^*(s)$ distribution for fitting sedimentation data with a distribution of sedimentation boundaries of non-diffusing particles, such as shown in Fig. 2.3. *Panel A*: Simulated sedimentation profiles of a normal distribution of large particles at 8000 rpm, shown in time intervals of 150 sec (black lines). The best-fit ls-$g^*(s)$ distribution was computed with a very coarse grid of only 25 s-values with $s_{min} = 1000$ S and $s_{max} = 2000$ S. The sequence of sequentially faster sedimenting step functions produces a stair case-shaped approximation of the boundaries (colored lines). Each value of ls-$g^*(s)$ corresponds to the height of a particular step, for example, the steps indicated by the blue, green, and magenta arrows. The initial height of each step — i.e., the ls-$g^*(s)$ values — is adjusted by a least-squares criterion to match the smooth experimental data. *Panel B*: Shown is the best-fit ls-$g^*(s)$ distribution (black), with colored arrows highlighting the particular values from the corresponding steps in A.

3.2.5 Transformation to Standard Conditions

When calculating the sedimentation coefficient and other distributions, it is prefer-able not to immediately transform the sedimentation coefficients to standard con-ditions of water at 20°C. At first, at least, it is advantageous to maintain the scale of experimental s-values, because it is close to the experimental observation and does not yet involve an additional interpretative step.

Frequently, the distribution exhibits many peaks arising from several different sedimenting components potentially with different partial-specific volume. Since the transformation to standard values via Eq. (1.22) requires knowledge of the partial-specific volume $\bar{v}$, it can take place only after the identity of the peaks has been established, and then it is strictly correct only for the peak associated with the particular $\bar{v}$-value, with the reliability of the remaining peak $s_{20,w}$-values being more uncertain. For this reason, a better approach for systems showing heterogeneity is to restrict the transformation to individual peaks.

For the calculation of some types of distributions, frequently an operationally defined $\bar{v}$ must be used to evaluate the kernel functions. Once the assignment of the species identity and true $\bar{v}$-value has taken place, corrections of the peak ap-parent molar mass values can be made to arrive at the true molar mass values, and similarly, the transformation of the frictional ratio to the true physical scale can be carried out, as discussed in Section 1.2.1 and Part I, Chapter 2.

> The `Options`▷`Size Distributions` menu in `SEDFIT` offers several transformation func-tions for sedimentation coefficient distributions after they are fitted to the raw data. The `Calculator` menu offers a function to transform integrated s-values.

A different situation arises in global modeling of SV data acquired under dif-ferent solvent conditions. In order to arrive at a single sedimentation coefficient distribution, the distribution s-values and all other molecular parameters need to be kept in standard units of water at 20°C, with the transformation into the dif-ferent experimental data spaces becoming part of the model [65].

3.3 REGULARIZATION

It is in the nature of the Fredholm integral equations such as Eq. (3.1) and the discretized form Eq. (3.5), that when using a sufficiently fine discretization with very large N, the resulting distribution will become very sensitive to the experimental noise, or ill-conditioned. In fact, in the extreme case, if unchecked, data noise will be amplified into a series of spikes that tend to dominate the distribution.[7] Limitations from ill-posed integral equations are well known in other biophysical disciplines, for example, dynamic light scattering [56, 122], electron spin resonance [123], NMR [124], and similarly, regularization suppressing noise amplification is wide spread in image analysis and tomography [125–127] and has applications in surface plasmon resonance [128], chemical kinetics [129,130], and protein folding [131]. It is obvious that we would not expect unlimited resolution from experimental data, and it is important to be aware of this fact and address it in the numerical analysis.

In the context of AUC, the detailed origin of this limitation can be visualized best, again, with the ls-$g^*(s)$ distribution of non-diffusing particles. At a rough discretization, as illustrated in Fig. 3.5, the steps are large and their height is unambiguously defined by the requirement to deviate as little from the data as

---

[7]Such unlimited error propagation is analytically predicted for Fredholm integral equations of the type Eq. (3.1) by the lemma of Riemann–Lebesgue, which states that [56]

$$\lim_{\omega \to \infty} \int_a^b \chi(s) \sin(\omega s) ds = 0 \,,$$

i.e., for even the smallest signal contribution σ, such as from noise, there is an oscillation frequency ω that can produce an arbitrarily large oscillation on the distribution $\gamma(s)' = \gamma(s) + \sin(\omega s)$ not changing the model more than σ. Roughly speaking, because the integral over a high-frequency oscillation (relative to the resolution of the experiment) vanishes, within the finite noise of the data, we cannot decide between distributions with and without such oscillations. Finer points need to be made with regard to the numerically discretized problem, and to account for the exclusion of negative concentrations, but mathematically this is at the root of the difficulty of calculating distributions.

To illustrate the discretized problem in Eq. (3.5), we can make a gedankenexperiment where we imagine a particular best-fit distribution obtained with a given grid of size N, e.g., the black one sketched here,

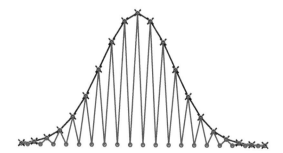

producing a satisfactory fit to the data (not shown) with a certain rmsd. We can then imagine a distribution at twice the grid size $N^* = 2N$, encompassing the old grid (black crosses) but with additional new grid points inserted between the previous grid points (red circles). At the matching grid points we assign to the new distribution the same values as the old distribution, and we fix the values at the new intermediate grid points to zero. The result looks like the oscillating red curve in the sketch. The key here is that since we have not changed anything in the summation of the model functions, just added zero amplitude terms, the rmsd will be exactly identical for both distributions. This holds true for any distribution model: Additional zero amplitude terms do not contribute anything to the fit, but can introduce wild oscillations in the distribution. Clearly there is ambiguity in the result, and we cannot just take any naïvely calculated distribution at face value and interpret as molecular properties.

possible. Unfortunately, the result may not be satisfactory for small N because the model is too coarse. With finer discretization, the length and height of each step decreases and the staircase approximation of the experimental data resembles a smoother function. However, when the resolution exceeds a certain value, $N \gg N^*$, the amplitudes of each step are smaller than the noise in the data.[8] At this point, the detailed shape of the best-fit distribution can translate and magnify the random noise of the experimental data in the distribution (blue trace in Fig. 3.6). Further, many different distributions, i.e., combination of step heights, exist that fit the data statistically indistinguishably well.

This problem needs to be addressed with the computational technique of regularization. Regularization provides the most parsimonious distribution function that is consistent with the experimental data. This is achieved by gently imposing a relationship among the $\gamma(s_l)$, which may be mathematically expressed with a functional $\Omega[\gamma(s_l)]$, such that oscillations are suppressed and a parsimonious distribution $\gamma(s_l)$ is obtained, but only to the extent that the quality of fit does not change significantly. Instead of solving Eq. (3.5), we solve

$$\operatorname*{Min}_{\gamma(s)\geq 0} \left(\alpha\Omega[\gamma(s)] + \sum_{i,j} \left[a(r_i, t_j) - \sum_{l=1}^{N} \gamma(s_l)\chi_1(s_l, r_i, t_j)\Delta s_l - b(r_i) - \beta(t_j) \right]^2 \right)$$

(3.6)

where $\Omega[\gamma(s_l)]$ expresses a penalty for non-parsimonious $\gamma(s_l)$ and α is a scaling parameter. Any positive value of α will create bias in the minimization that causes a deviation of the distribution from its overall best-fit value. Correspondingly, the χ^2 value of the fit is elevated. The scaling parameter can be iteratively adjusted such that

$$\frac{\chi^2(\alpha)}{\chi^2(\alpha = 0)} = F(\nu_1, \nu_2, 1 - P),$$

(3.7)

with $F(\nu_1, \nu_2, 1 - P)$ denoting the critical value of the F-statistics based on the confidence level P and degrees of freedom ν [72,132]. In this way it can be ensured that the regularization does not apply significant constraints to the fit. For example, the magenta distribution in Fig. 3.6 provides very gradual changes in the consecutive $\gamma(s_l)$-values, yet has a statistically indistinguishable quality of fit from the naïve overall best-fit shown in blue.[9]

---

[8]N^* is the minimal resolution that is required to model the boundary within the noise of the data acquisition. It may be derived from the maximum slope of the boundary da/dr and the noise of data acquisition δ, using the idea that the discretization Δs should lead to a radial spacing Δr of non-diffusing particles that satisfies $da/dr \leq \delta/\Delta r$, from which follows $\Delta s \leq \delta/(\omega^2 tr(da/dr))$, or

$$N^* = \frac{(s_{\max} - s_{\min})\omega^2}{\delta} \times \max\left[rt\frac{da(r,t)}{dr} \right].$$

[9]Regularization must not be confused with smoothing: If a smoothing operation were to be applied to the distribution, then there would be no control over the quality of fit of the resulting smoothed distribution relative to the raw data. At best, this would sacrifice detail that could be

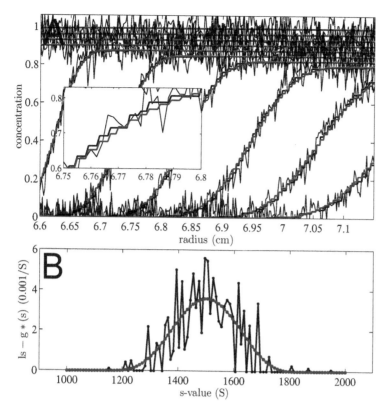

Figure 3.6 Effect of regularization illustrated with the ls-$g^*(s)$ distribution fit to noisy data. The example is based on the same sedimentation data as shown in Fig. 3.5, but with added noise, and using a distribution with a resolution of 100 grid points between s_{min} and s_{max}. *Panel A:* Shown here is a subset of the data (black noisy lines) along with the best-fit step functions corresponding to ls-$g^*(s)$ without regularization (blue step functions) and with Tikhonov regularization at P = 0.68 (magenta step functions). Both fit the data very well. However, zooming into a particular section of a scan in the inset, it can be discerned that the regularized distribution has even step sizes, whereas the step sizes of the un-regularized naïve distribution are seemingly random and very poorly constrained by the bounds of the noisy data. *Panel B:* The ls-$g^*(s)$ distributions with (magenta) and without (blue) regularization that correspond to the step function fits in *A*.

Finally, the choice of regularization functional Ω can embed certain ideas about the sample. For example, macromolecules with a smooth size distribution can be described with Tikhonov–Phillips regularization where [14, 56, 133, 134]

$$\Omega_{\mathrm{TP}}[c] = \int (dc/dx)^2 dx \,. \tag{3.8}$$

By contrast, for distributions of particles where a few discrete species are expected, the information entropy

$$\Omega_{\mathrm{ME}}[c] = \int c(x) \log c(x) dx \tag{3.9}$$

---

reliably interpreted, but smoothing also might create misleading features that are not warranted by the data. There would be no obvious rational criterion for the amount of smoothing.

is advantageous [14,135,136]. Different expressions based on specific prior knowledge in a Bayesian approach are discussed in Section 5.7. The concept of regularization is illustrated in Fig. 3.7, where from all distributions that fit the data indistinguishably well, we can select the one that most conforms with our prior notion of parsimony, following Occam's razor. This honors the full information content of the data, but avoids misleading details that are statistically not warranted.

The lower the signal/noise ratio in the experimental data, the lower their information content and the more the distribution is governed by minimum values

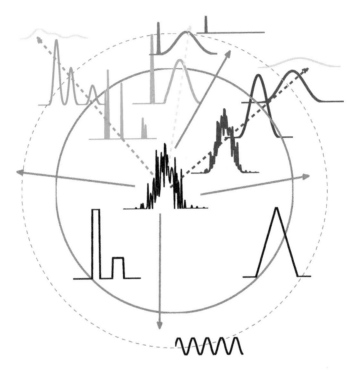

Figure 3.7 Cartoon depicting the choice and scaling of regularization in the function space of possible distributions. The blue noisy distribution in the center shall represent the overall best-fit distribution with fit quality of χ_0^2. All other functions also represent possible distributions, and they are arranged in concentric circles with increasing radii representing increasing χ^2 of the fit to the experimental data (arrows). The χ^2 values can be scaled using F-statistics to a P-value, and the circles are drawn for $P = 0.68$ (solid circle) and P = 0.95 (dashed circle). Thus, all functions within the solid circle fit the data statistically indistinguishably well on a confidence level of one standard deviation. The direction of the red dashed arrow with distributions colored in purple, magenta, red, and orange are functions arranged with increasing value of $-\int (dc/dx)^2 dx$ as in Tikhonov regularization, whereas along the direction of the green dashed arrow the cyan, green, and yellow distributions have increasing entropy $-\int c(x) \log c(x) dx$. The light blue dashed arrow points in the direction of increasing weighted entropy with prior assumption of the distribution consisting of a delta function at a particular s-value. All colored directions correspond to increasing parsimony, and at equivalent radii they fit the data equally well, but they are based on different prior expectations that embed what we know about the sample. The distribution solely resembling the prior expectation will be located farthest from the center.

of the regularization functional. With the standard forms Eqs. (3.8) and (3.9), low signal/noise will create very broad features, and ultimately a flat distribution. With Bayesian priors the distribution will ultimately just be the prior. *Vice versa*, data with higher signal/noise ratio will have higher information content, and generate higher penalties for featureless distributions because of their stronger misfit, which results in sharper distributions.

The selection of Tikhonov and maximum entropy regularization, or both, can be made in the SEDFIT menu Options ▷ Size Distributions. The *P*-value will be entered in the respective distribution model parameter box.

3.4 DISTRIBUTIONS AND s_w IN THE TRANSPORT METHOD

All differential sedimentation coefficient distributions of the above structure have in common a unique relationship to the weighted-average sedimentation coefficient s_w. We will demonstrate this for the one-dimensional case. Assuming the distribution provides a good fit to the experimental data with regard to the plateau signal and the temporal derivative of the integral, and after removal of best-fit systematic baseline offsets, we can insert the definition of the distribution $\gamma(s)$ (Eq. (3.1)) to replace the data in the definition of s_w (Eq. (2.45)):

$$s_w = \lim_{t \to 0} \left[\frac{1}{\omega^2 r'^2 \int_{s_{\min}}^{s_{\max}} \gamma(s)\chi_1(s,r',t)ds} \left(-\frac{d}{dt} \int_m^{r'} \int_{s_{\min}}^{s_{\max}} \gamma(s)\chi_1(s,r,t)dsrdr \right) \right].$$
(3.10)

Exchanging the order of the integral, we have

$$s_w = \lim_{t \to 0} \left[-\frac{1}{\omega^2 r'^2 \int_{s_{\min}}^{s_{\max}} \gamma(s)\chi_1(s,r',t)ds} \left(\int_{s_{\min}}^{s_{\max}} \gamma(s) \int_m^{r'} \frac{d}{dt}\chi_1(s,r,t)rdrds \right) \right].$$
(3.11)

For the temporal derivative $d\chi/dt$, we can insert the Lamm equation (2.30) and execute the radial integration. Analogous to the derivation of the transport method, Eq. (2.39), due to the vanishing diffusion fluxes in the plateau the only remaining term is

$$s_w = \lim_{t \to 0} \left[-\frac{1}{\omega^2 r'^2 \int_{s_{\min}}^{s_{\max}} \gamma(s)\chi_1(s,r',t)ds} \int_{s_{\min}}^{s_{\max}} \gamma(s)\left(-\chi_1(s,r,t)s\omega^2 r'^2 \right) ds \right]$$

$$= \lim_{t \to 0} \left[\frac{1}{\int_{s_{\min}}^{s_{\max}} \gamma(s)\chi_1(s,r',t)ds} \int_{s_{\min}}^{s_{\max}} \gamma(s)s\chi_1(s,r,t)sds \right].$$
(3.12)

Since $\lim_{t \to 0} \chi_1(s, r, t) = 1$, i.e., the plateau concentrations of the normalized Lamm equation solutions are initially equal to the unit loading concentration, we have

$$s_w = \frac{\int_{s_{\min}}^{s_{\max}} \gamma(s) s \, ds}{\int_{s_{\min}}^{s_{\max}} \gamma(s) \, ds} \qquad (3.13)$$

or, after consideration of potentially different signal contributions from different species, we have

$$s_{w,\lambda} = \frac{\int_{s_{\min}}^{s_{\max}} \varepsilon_\lambda(s) \gamma(s) s \, ds}{\int_{s_{\min}}^{s_{\max}} \varepsilon_\lambda(s) \gamma(s) \, ds} \, . \qquad (3.14)$$

This is a natural extension of Eq. (2.46) for the continuous case.

This result could reasonably be used as a starting point and definition for s_w, but in doing so the foundation of s_w in the transport method would be missed. This provided the critical insight that for an integral of the sedimentation coefficient distribution to reflect the correct s_w of the sedimentation process, it is essential that the distribution describes the raw data $a'(r, t)$ well — specifically with respect to the signal magnitude in the plateau, and with respect to the radial integral over the sedimentation boundaries — but that it does not depend on the Lamm equation kernel to be the correct physical sedimentation model! This will play an important role in the analysis of interacting systems.

On the other hand, if these conditions are fulfilled, it makes sense to extend the definition of s_w using Eqs. (3.13)/(3.14) to situations where no distinct plateau can be identified in $a'(r, t)$, for example, due to the presence of small species with strong back-diffusion. Similarly, Eqs. (3.13)/(3.14) can be used in conjunction with integration limits $s_{\min}$ and $s_{\max}$ that correspond only to a subset of all sedimenting species, for example, describing only one of many discernable sedimentation boundaries. In the absence of plateaus it may be more difficult to rationalize integration limits, and the result may become dependent, to some extent, on that the underlying sedimentation model is correct.

SEDFIT offers an integration tool to determine s_w. The limits $s_{\min}$ and $s_{\max}$ can be specified either in the graphics by using the mouse, or, after setting a flag in the Options ▷ Size Distributions menu, they can be read from entries of a user-supplied two-column ASCII file containing one or more rows with $s_{\min}/s_{\max}$ pairs.

3.5   STATISTICAL CONFIDENCE LIMITS

How to assign confidence limits to the resulting distribution is an important question when we interpret the results. A powerful tool is an exhaustive series of Monte-Carlo simulations, in which the experiment is repeated *in silico* many times (typically on the order of $\sim$1000) with different batches of synthetic statistical noise.

This can often be done fairly quickly, since a significant part of the computation does not need to be repeated at each iteration [137] (Appendix B). A statistics can be built of the histogram of resulting distributions, yielding a whole probability function at each s-value for certain distribution values $\gamma(s_l)$ to occur.

This is illustrated in Fig. 3.8, which is based on a simulation of two species boundaries with relatively poor signal/noise ratio, with the total signal being 10-fold the noise and the second species at loading concentration equal to the level of noise. In 1000 iterations, each simulated data set was analyzed with diffusion-deconvoluted sedimentation coefficient distribution $c(s)$. Without regularization, the $c(s)$ distributions exhibit sharp and sometimes fragmented peaks, but different iterations scatter significantly with respect to their precise peak location and peak substructure. This is a result of error amplification typical for Fredholm integral equations.

From the histogram of amplitudes at each s-value, one can determine the limits of the central 68% of amplitude values, and then connect these limits from different s-values to form contour lines representing the upper and lower confidence limits of the distribution. However, this strategy for the error analysis of the distribution has significant drawbacks: Because the detailed peak locations are variable, and their sharpness is limited only by the resolution N, the lower confidence limits will usually be close to zero. Furthermore, this strategy of statistical analysis does not take advantage of a strong correlation between heights and locations of peaks and sub-peaks; in fact, a distribution completely resembling a probability contour line would lead to a bad fit.

Usually, more important than the statistical confidence limits of the distribution itself, are errors of quantities based on the integrals over the distribution, such as $\int \gamma(s)ds$ and s_w-values. These are much better constrained by the experimental data, with the former reflecting the boundary amplitudes and the latter the transport (change in area) with time. Correlated, compensatory oscillations between neighboring s-values will not impact the result of the integration, depending on the choice of the integration limits.

As illustrated in the insets in Fig. 3.8B, if the respective integrals are evaluated for all iterations, the resulting frequency histograms are remarkably narrow. For loading concentrations, the standard deviation is approximately a factor ten smaller than the amplitude of noise in an individual data point (and the error in the total concentration is further reduced by a factor $\sim$2). This can be attributed to the large number of data points reporting on the boundary heights. The s_w histograms lead to a $\sim$0.3% error for the more abundant species, and an approximately 10-fold higher error ($\sim$2%) for the species with loading concentration equal to the noise in data acquisition.

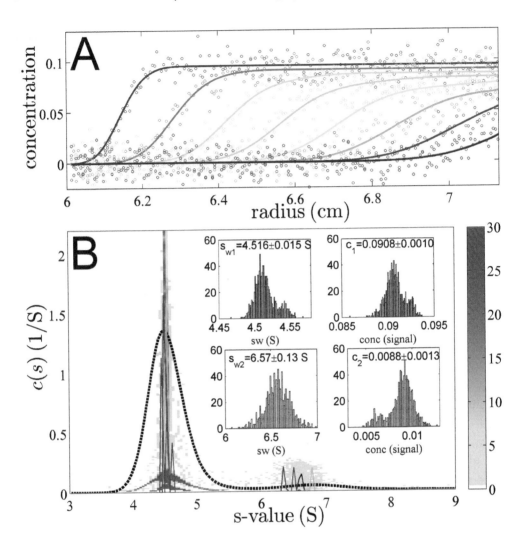

Figure 3.8 Noise propagation in sedimentation coefficient distribution analysis. *Panel A*: Sedimentation velocity boundaries at 50,000 rpm of two species, a 65 kDa, 4.5 S species at a loading concentration of 0.09 and a 130 kDa, 6.5 S species at a loading concentration of 0.01, in the presence of 0.01 signal units of normally distributed noise. Only every 6th scan is shown, from a total of 50 scans at 300-sec intervals, each with a 0.003-cm radial data interval, comprising a total number of 18,700 data points (circles). The solid lines are the best-fit boundaries from $c(s)$ analysis. *Panel B*: To illustrate the noise propagation, the simulation was repeated 1000 times with different draws of random noise, and a $c(s)$ sedimentation coefficient distribution analysis on a grid of 300 s-values between 1 S and 12 S was performed on each, without regularization. Shown vertically at each s-value is a histogram of $c(s)$ amplitudes for the particular s-value, with the frequency encoded in the cyan-to-magenta color scheme. As illustration of the source of scatter, $c(s)$ distributions of four particular instances of the simulation are shown (solid thin red, blue, black, and green lines). Also, for comparison, the distribution calculated with maximum entropy regularization at P = 0.68 is depicted as a bold dotted line (in 10-fold magnification). While the individual distributions have a lot of scatter, the peaks are strongly correlated. The insets show the propagation of noise into the integrated properties across the region of the first peak in the range of 3.5 S to 5.5 S, and the second peak in the range from 5.5 S to 7.5 S: Histograms of s_w-values are in the left column of insets and calculated boundary amplitudes on the right, for the first peak (top row) and the second peak (bottom row).

The Statistics menu functions in SEDFIT include Monte-Carlo analysis of the distribution both for the purpose of calculating confidence contour lines, as well as the statistical errors on the integrated loading concentrations and s_w over one or more integration ranges.

Often the distributions will include additional non-linear parameters, such as frictional parameters and/or meniscus positions. These can contribute to the errors in loading concentrations and weighted-average s-values. In this case, an estimate of the overall errors may be found efficiently by determining the upper and lower statistical confidence limits of the non-linear parameters with F-statistics, and then carrying out distribution error analyses with the non-linear parameter fixed at either limit. More on practical error analysis can be found in Section 8.3.

CHAPTER 4

Distributions of Non-Diffusing Particles

B EFORE approaching the general problem of resolving particle heterogeneity and diffusion, we consider the simpler problem of a distribution of very large, virtually non-diffusing particles. This provides a starting point for the practice of distribution analyses that is conceptually easier, links up with historical methods, and can be generalized later in various ways to problems exhibiting diffusion.

Although distributions of non-diffusing particles are strictly not possible since diffusion is always present to some extent, they are useful in practice whenever SV experiments are conducted with very large, fast sedimenting particles. For these systems, the limiting case of non-diffusing particles is justified by the very short experiment providing little diffusion time, in addition to the intrinsically very small diffusion coefficient of large particles. In other words, the transport fluxes from diffusion are much smaller than those from sedimentation and negligible in comparison. Examples include cell suspensions [138], amyloid fibrils [139], some large polydisperse multi-protein complexes or viral particles [140], polysaccharides [141], and nano-particles and dispersions [96, 97, 121, 142, 143]. In addition, this can be a useful approximate analysis tool where diffusional deconvolution is not successful or not desired.

The relationship between the distribution shape and the boundary shapes is illustrated in Fig. 4.1. Whereas for a single non-diffusing species we saw only step functions with infinitely sharp boundaries, for continuous distributions of non-diffusing particles we have boundaries with continuous slopes that directly reflect the shape of the distribution. Later we will work out in more mathematical detail how the distribution governs the radial and temporal derivative. But from a cursory inspection of Panel A of Fig. 4.1, smooth distributions of non-diffusing particles can provide families of concentration profiles that may superficially look quite similar to those of one or a few diffusing species (e.g., those in Fig. 3.2C), and if only a small time interval is considered, one may be misled to naïvely model the data as one or more strongly diffusing species. However, the rapid boundary broadening of

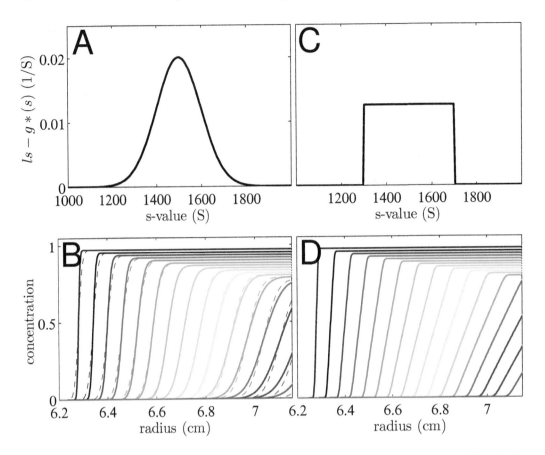

Figure 4.1 Continuous model distributions of large particles and their sedimentation boundaries at 8000 rpm shown in time intervals of 180 sec. Due to their rapid sedimentation, transport from diffusion is negligible. *Panel A* and *B*: Normally distributed particles. What may appear at first sight to be diffusion broadening of the boundaries, is a reflection of the particle distribution. The best-fit impostor single-species Lamm equation solution of a diffusing species is shown as dashed lines. *Panels C* and *D*: The equivalent case for a rectangular shaped particle size distribution.

the non-diffusing polydisperse distribution would lead to apparent diffusion coefficients impossibly large for particles of the given sedimentation rate. We have seen this above as a tell-tale sign of unrecognized polydispersity. However, the initially very steep and later rapidly spreading boundaries in Fig. 4.1A cannot be modeled by diffusion if the entire time course is considered (similar to the fit of the heterogeneous diffusing sample with an impostor single-species model in Fig. 3.3.) The absence of diffusion is easier to discern from the rectangular model distribution in Fig. 4.1C/D. Unfortunately, in practice it seems many particle preparations are indeed more smoothly distributed, often close to the Gaussian size distribution illustrated in Fig. 4.1A/B.

4.1 THE APPARENT SEDIMENTATION COEFFICIENT DISTRIBUTION ls-$g^*(s)$

Let us denote as $\chi_{\text{nd},1}$ the migrating step function of a single, non-diffusing species sedimenting at unit loading concentration with sedimentation coefficient s, following Eq. (2.10). In analogy to the general one-dimensional sedimentation coefficient distribution described in Eq. (3.1), we can then write the definition of the apparent sedimentation coefficient distribution ls-$g^*(s)$, or short $g^*(s)$, as the integral equation [121]

$$a(r,t) \cong \int_{s_{\min}}^{s_{\max}} g^*(s)\chi_{\text{nd},1}(s,r,t)ds + b(r) + \beta(t). \qquad (4.1)$$

The designation "ls" arises from the least-squares boundary modeling involved, and the asterisk indicates that this is only an "apparent" sedimentation coefficient distribution, since it is based on the approximation that diffusion is absent. However, for very large particles, ls-$g^*(s)$ will be very close to the true sedimentation coefficient distribution, which we have termed $c(s)$ and will be described below.[1] Because the equations are all linear and can be solved by linear algebra methods, including the Tikhonov regularization, the computational effort is relatively small and straightforward (Appendix B). The conceptual simplicity of this approach highlights the power of its foundation of directly modeling the experimental data in their original data space with explicit models.

Once the range of s-values is discretized, this distribution will produce sums of step functions, each migrating at a different rate s_l, such that stair case-shaped model functions for the concentration profiles are generated. As mentioned above and shown in Fig. 3.5, the best-fit height of each sequential step directly reflects the amplitude of the sequential values of ls-$g^*(s)$. If the underlying particles are truly non-interacting and ideally sedimenting with negligible diffusion, a virtually perfect fit — strictly within the noise of the data acquisition — will be achieved by ls-$g^*(s)$ across the entire range of recorded sedimentation data.[2] This is illustrated in Fig. 4.2 for a suspension of single-cell organisms (under starving conditions leading to the cessation of active swimming) [138].

A variation can be useful for some protein systems where, in addition to the large non-diffusing species, a small diffusing species is present in significant amount in the sample. This case can often be described well with the combination of a

---

[1]A variation of Eq. (4.1) describing the case of analytical electrophoresis can be found in the SEDFIT models menu Other Geometries. In this case, particle migration is with constant velocity due to a constant parallel force field, and no radial dilution occurs. A different variation of Eq. (4.1) has been proposed by Walter and colleagues [144] for centrifugal analysis, with kernels $\chi_{\text{nd},1}$ that lack radial dilution in order to accommodate parallel-walled sample containers. However, the justification of this model is questionable on physical grounds (see Part I, Section 5.2.1.1).

[2]Due to the discrete nature of the fitting function, at coarse discretization levels the residuals can exhibit characteristic saw-tooth patterns reflecting the mismatch between the discrete step functions and the smoothly sloping data.

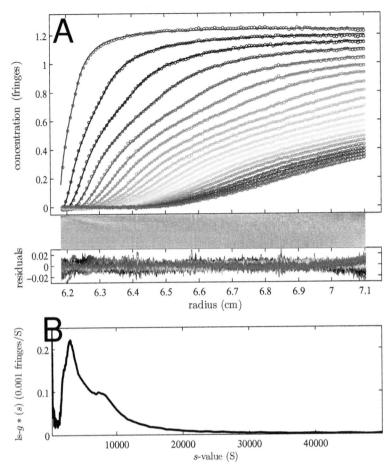

Figure 4.2 Distribution model ls-$g^*(s)$ applied to *Spiroplasma melliferum* BC3 [138]. *Panel A*: Data (circles), fit (solid line), and residuals overlay and bitmap of interference fringe profiles recorded at 3000 rpm in intervals of ~30 sec. The distribution was discretized with $s_{min} = 100$ S and $s_{max} = 100,000$ S, with $N = 400$ on a logarithmic scale, using Tikhonov regularization with $P = 0.68$. After non-linear regression of the meniscus position, the best-fit rmsd of the fit is 0.0044 fringes. For clarity, only every 10th data point of every 5th scan is shown. *Panel B*: ls-$g^*(s)$ distribution corresponding to the fit in A.

single, discrete diffusing species with the ls-$g^*(s)$ distribution

$$a(r,t) \cong c_1 \chi_1(s_1, D_1, r, t) + \int\limits_{s_{min}}^{s_{max}} g^*(s) \chi_{nd,1}(s, r, t) ds + b(r) + \beta(t), \qquad (4.2)$$

where for the small species, at loading concentration c_1, sedimentation follows the regular Lamm equation solution Eq. (2.30) with s_1 and D_1. Similar to the example in Fig. 3.2D, migration from diffusion of the small species may be comparable to the migration from sedimentation of the larger species. This model prevents the large diffusional spread of the small species to interfere and bias the distribution of large species (see, e.g., the application of this concept to a large polydisperse mixture of protein complex in the presence of a small receptor domain [140]).[3]

Both the standard ls-$g^*(s)$ model and the extension with an extra discrete species can be invoked in the Model menu in SEDFIT. Because typically diffusional broadening of the apparent sedimentation coefficient distribution is to be expected, the default regularization is switched to Tikhonov–Phillips regularization, which performs better for broad and smooth distributions.

It is easily possible to extend the ls-$g^*(s)$ model to compressible solvents by exchanging the kernel in Eq. (4.1) from the ideally sedimenting non-diffusing sedimentation model to the sedimentation profiles given by Eqs. (2.18) and (2.19) for a non-diffusing particle in a compressible solvent. In this way, the resulting ls-$g^*(s)$ model will inherit full consideration of the solvent compressibility, without introducing any further approximations or computational complications. Similarly, effects of finite scan speed can be accounted for in the ls-$g^*(s)$ distribution when using the kernel Eq. (2.28).

Unfortunately, a similar extension is not directly possible for particle sedimentation in density gradients dynamically established from high concentrations of sedimenting co-solutes, because for this case no analytically closed form for the theoretical sedimentation profiles of a single, non-diffusing particle exists. However, this situation can be approximated by using suitable models solving the full Lamm equation with a very small diffusion coefficient,[4] an approach that can be implemented conveniently by using the $c(s)$ sedimentation coefficient distribution (Section 5.4) fixing the average frictional ratio to a large value.

The reader will note the seemingly circular approach to use full Lamm equation solutions as a tool for approximating the sedimentation of non-diffusing particles, which themselves are idealizations neglecting existing diffusion of these particles. In fact, one may question outright the need of modeling SV data with the approximation of non-diffusing particles, considering that modern algorithms for solving the Lamm equation are sufficiently fast and precise not to represent any real hurdle. However, there are several reasons for why in some cases one may prefer an apparent sedimentation coefficient distribution over the distribution models incorporating diffusion, provided a good fit is achieved in either model: These include the

---

[3]Optically unmatched buffer salts can sometimes contribute significantly to the signal, as described in detail in [47] and discussed in Part I, Section 4.3.4. Frequently, the study of very large particles is carried out at rather low rotor speeds, such that moderate amounts of optically unmatched buffer salts in studies at very low rotor speeds may not need to be explicitly considered in the analysis, as pointed out by Berkowitz and Philo [145]. However, high concentrations of unmatched buffer co-solutes, such as 10% glycerol in the sample buffer with water as optical reference, can generate measurable signals even at very low rotor speed [145].

[4]Most algorithms for the numerical solution of the full Lamm equation including diffusion terms exhibit numerical instabilities when D is set to zero (Appendix A).

model-independence of the weighted-average s_w-value, the opportunity to extract sedimentation behavior with fewer parameters, and the relationships between $g^*(s)$ and the boundary shape.

4.2 REPRESENTATION OF THE SEDIMENTATION BOUNDARY SHAPE

Among the most compelling arguments for $g^*(s)$ is the attempt to simplify the analysis for studies where a more appropriate description cannot be found. This includes, for example, the study of distributions of highly concentrated polymers where hydrodynamic and thermodynamic non-ideality becomes a dominant factor governing the boundary shape. Even though accurate theoretical models for sedimentation and diffusion of such systems are still an active area of research, a quantitative empirical description of the average s-value under experimental conditions can already be achieved with ls-$g^*(s)$. This rests on the transport method only requiring the integral of the boundaries to be evaluated accurately, irrespective of the details of the model (see above). Similarly, $g^*(s)$ distributions are very useful in the study of systems of very large particles where diffusion is well understood but experimentally not well captured in SV, such that apparent distributions are more commensurate to the SV data.

An attractive feature of ls-$g^*(s)$, and $g^*(s)$ in general, is that it provides a quantitative description of the shape of the sedimentation boundary at a particular point in time t_1 in units of apparent sedimentation coefficients s^*. The following considerations require us to focus only on the signal from the sedimenting particles, $a'(r, t_1) = a(r, t_1) - b(r) - \beta(t_1)$, in the absence of systematic noise contributions. If we can describe the boundary shapes $a'(r, t_1)$ through $g^*(s)$, this defers all complexity of the sedimentation process to the inspection of the *apparent* sedimentation coefficient distribution. Even though this diverges from the original goal of direct boundary modeling, the change in units by going from $a'(r, t_1)$ to $g^*(s, t_1)$ may help to suggest hypotheses for the physical process.

For ideal sedimentation processes, the relationship between $g^*(s, t_1)$ and properties of the sedimenting particles is straightforward. The spatial derivative of the boundary

$$\frac{da'(r, t_1)}{dr} = \frac{d}{dr} \int_{s_{\min}}^{s_{\max}} g^*(s, t_1) \chi_{\mathrm{nd},1}(s, r, t_1) ds \tag{4.3}$$

can be computed *via* the radial derivative of the step functions $\chi_{\mathrm{nd},1}(s, r, t_1)$ defined in Eq. (2.10),

$$\frac{d\chi_{\mathrm{nd},1}(s, r, t_1)}{dr} = e^{-2s\omega^2 t_1} \delta(r - me^{\omega^2 s t_1}). \tag{4.4}$$

The transformation

$$s^* = \frac{1}{\omega^2 t} \log\left(\frac{r^*}{m}\right) \tag{4.5}$$

allows us to express the radial coordinate *via* the sedimentation coefficient s^* of an ideally sedimenting, non-diffusing particle that, after starting at the meniscus and migration for the time $t = t_1$, arrives at a particular radius r^*. Accordingly, we can re-write the δ-function in Eq. (4.4) as

$$\delta(r^* - me^{\omega^2 st}) = \frac{e^{-\omega^2 st}}{m\omega^2 t}\delta(s - s^*),\tag{4.6}$$

insert these results into Eq. (4.3), and obtain after integration the Bridgman relationship[5]

$$g^*(s^*, t_1) = \omega^2 t_1 \times \frac{r^{*3}}{m^2} \times \frac{da'(r^*, t_1)}{dr},\tag{4.7}$$

with the asterisk denoting the fact that this is an approximation based on the relationship between s^* and $r^*(t)$ of a non-diffusing particle Eq (4.5). For clarity, we have explicitly marked the time dependence of $g^*(s^*)$. Eq. (4.7) states that $g^*(s^*, t_1)$, when derived from a single radial profile at a particular time point, can be regarded as a direct representation (or transformation) of the shape of the sedimentation boundary.[6]

For non-diffusing particles, all features of the sedimentation coefficient distribution will be reflected directly in the radial derivative of the macromolecular signal. However, all other determinants of the sedimentation process that influence the boundary shape, such as diffusional broadening or boundary sharpening from hydrodynamic non-ideality, will also produce corresponding features in our estimate of $g^*(s^*, t_1)$, possibly generating a complex evolution with time.

---

[5]This was first reported in 1942 by Wylbur Bridgman and applied to the study of the size distribution of glycogen [146]. The relationship was particularly useful in conjunction with the Schlieren optical system, which reports directly on the concentration gradients.

[6]It is instructive to consider the relationship between ls-$g^*(s)$ and the boundary shape in the hypothetical case of rectangular geometry and constant migration velocity v. In this case, the boundary steps for a non-diffusing species would simply be the Heaviside step function $H(r - vt)$, and in the noise-free case, any momentary concentration gradient $a(r, t_1)$ can be exactly described as a velocity distribution $g^\circ(v)$ with

$$a(r, t_1) = \int g^\circ(v, t_1)H(r - vt_1)dv.$$

As pointed out previously, since $\left(\partial/\partial r\right)H(r - vt_1) = \delta(r - vt_1)$, with the transformation $r = vt_1$ it is $\delta(r - vt_1) = t_1^{-1}\delta(v - v^*)$ and the radial derivative of the concentration gradient becomes

$$\frac{da(r, t_1)}{dr} = \int g^\circ(v, t_1)t_1^{-1}\delta(v - v^*)dv = t_1^{-1}g^\circ(\frac{r}{t_1}, t_1),$$

i.e., the velocity distribution $g^\circ(v)$ is identical to the radial derivative of the concentration profile *modulo* a temporal normalization factor. Thus the distribution ls-$g^*(s)$ is deeply connected to the radial derivative, with the sole modifications being the geometrical factors for sector-shaped cells, and the practical adaptation to rephrase the identity to a least-squares fit. In this analogy, the need for regularization corresponds to the well-known problem of ill-determined derivatives of noisy discrete data. The distribution approach rephrases the differentiation to an integral equation, and the distribution analysis offers a natural way to integrate information from many samples at once.

A practical aspect of representing the sedimentation data in terms of $g^*(s^*, t_1)$ is that Eq. (4.3) above assumed it is possible to fit the data within the noise of data acquisition with the ls-$g^*(s)$ model to extract the boundary shape information. This poses a dilemma: If the time range of scans is very small — sufficient to approximate a single time point t_1 — then it will not be possible to achieve a reliable separation of sedimentation signal from both time-invariant and radial-invariant noise components $b(r)$ and $\beta(t_1)$, and the derivative will cause significant noise amplification. If, on the other hand, the experimental data considered span a sufficient time interval to accomplish reliable systematic noise composition, then a conflict arises with the assignment of a single time point t_1 to the distribution, and good fit cannot be achieved except for truly non-diffusing, ideally sedimenting particles.

4.3 $g^*(s)$ AND THE RATE OF CHANGE OF THE SEDIMENTATION SIGNAL

In historical applications of AUC to the study of dispersions, consisting of poly-disperse mixtures of very large particles, a custom-built stationary detector was used in several laboratories to record the change of signal with time at a fixed radius [96, 97, 147–149]. In this approach, the temporal derivative of the signal is related to the distribution of non-diffusing particles

$$\frac{da'(r_1, t)}{dt} = \frac{d}{dt} \int_{s_{\min}}^{s_{\max}} g^*(s)\chi_{nd,1}(s, r_1, t)ds \tag{4.8}$$

(with r_1 the radius of the detector). The temporal derivative of Eq. (2.10) is [150]

$$\frac{d\chi_{\mathrm{nd},1}(s, r_1, t)}{dt} = -2\omega^2 s \chi_{\mathrm{nd},1}(s, r_1, t) - m\omega^2 s e^{-\omega^2 st}\delta(r_1 - r_b(t)) \tag{4.9}$$

with $r_b(t)$ the boundary position of a suspension of non-diffusing particles as in Eq. (2.3). Notably, the time-derivative has two terms, due to the temporal dependence of the radial dilution, in addition to the migration of the boundary. This causes a signal change at the detector radius r_1 arising from the radial dilution of the plateau of all slowly sedimenting particles that have not migrated to the detector yet, superimposed by those signals from fast particles that just reach the detector. At a single time point, these cannot be distinguished.

The single-species signal Eq. (4.9) inserted into Eq. (4.8), generalized to any radius r and using the relationship Eq. (4.6) to transform the δ-function, leads to the integral equation

$$\frac{da'(r_1, t)}{dt} = -g^*(s^*)\frac{s^* e^{-2\omega^2 s^* t}}{t} - 2\omega^2 \int_0^{s^*} g^*(s')s' e^{-2\omega^2 s' t}ds', \tag{4.10}$$

where the change in integration limits arises from the Heaviside function in the integrant [151]. With the transformation between radius and sedimentation coefficient variable Eq. (4.5), a solution of the integral equation Eq. (4.10) is [151]

$$g^*(s^*) = -\frac{t}{s^*} \times \frac{r^{*2}}{m^2} \times \left(\frac{da'(r_1, t)}{dt} - \frac{2}{r^{*2}} \int_0^{r'=r^*(s^*)} \frac{da'(r', t)}{dt} r' dr' \right). \qquad (4.11)$$

This is the temporal analogue of the Bridgman equation Eq. (4.7).[7,8]

The essence of Eq. (4.11) for a single-radius detector is that the size distribution can be resolved unambiguously after taking into account the entire time course: Noting that for the smallest particles arriving last at the detector, the offset term from radial dilution of larger particles disappears (since all other particles are depleted depleted already), a correction for radial dilution can be applied to successively larger particles. This is the basis of the single-detector data analysis described by Scholtan and Lange [96].[9] In conjunction with absorbance and interference detectors with radial scanner or radial imaging, respectively, the implicit form of equation (4.10) was used in Stafford's dcdt program to iteratively calculate $g^*(s)$ [150], with the caveat that the time-derivative must be replaced in practice by a finite time difference of scans, as described in more detail in Section 6.2. For

---

[7]Briefly, this solution can be found by substituting $g^\circ(s) = g^*(s)se^{-2sw^2t_1}$ into Eq. (4.10). The result can be converted into a first-order linear differential equation by differentiation with respect to s^*

$$\frac{\partial g^\circ(s^*)}{\partial s^*} = -t_1 \frac{\partial^2 c(r(s^*), t_1)}{\partial s^* \partial t} - 2\omega^2 t_1 g^\circ(s^*),$$

which has the solution

$$g^\circ(s^*) = -t_1 e^{-2s^* \omega^2 t_1} \int_0^{s'=s^*} \frac{\partial^2 c(r(s'), t_1)}{\partial s' \partial t} e^{2s' \omega^2 t_1} ds'.$$

Back-substitution of g^* and integration by parts leads to Eq. (4.11) [151].

[8]It is instructive again to consider the $g^\circ(v)$ distribution and the temporal signal derivative at a specific radius r_1 in the hypothetical case of rectangular geometry and constant migration velocity v. In this case, the temporal derivative is

$$\frac{da(r_1, t)}{dt} = \int g^\circ(v) \frac{d}{dt} (H(r_1 - vt)) dv.$$

After transformation of differentiation variables $\left(\partial/\partial t \right) H(r_1 - vt) = -v \left(\partial/\partial r \right) H(r_1 - vt)$ the change into the velocity space leads to $-(v/t)\delta(v - r_1/t)$, and the temporal change of the boundary becomes

$$\frac{da(r_1, t)}{dt} = -\int g^\circ(v) vt^{-1} \delta(v - r_1/t) dv = -vt^{-1} g^\circ(\frac{r_1}{t}),$$

i.e., similar to the radial derivative, the velocity distribution $g^\circ(v)$ is identical to the temporal derivative of the concentration profile *modulo* a normalization factor. Obviously, the complications from radial dilutions do not arise, but otherwise there is a close correspondence with results in the sector-shaped geometry.

[9]An obvious practical disadvantage of the single fixed-radius detection is lower information content, chiefly the lack of opportunities to recognize systematic errors in the sedimentation model, for example, from diffusion or back-diffusion of small species, or RI noise contributions.

either detector configuration, the most direct approach would, of course, be a direct least-squares fit of the data $a'(r,t)$ with ls-$g^*(s)$. This maintains all information and minimizes noise amplification.

4.4 APPLICATION OF ls-$g^*(s)$ UNDER CONDITIONS WHERE DIFFUSION IS NOT NEGLIGIBLE

The special case of sedimentation of a single discrete ideally sedimenting and diffusing species will serve as an illustration for the application of ls-$g^*(s)$ to sedimentation processes more complex than that of non-diffusing particles. It is illustrated in Fig. 4.3 for the ideal case of noise-free data, excluding at first potential problems arising from systematic baseline offsets. The ls-$g^*(s)$ in Panel A is based on a narrow time interval to approximate an instantaneous analysis.

To examine this from a theoretical perspective, we can make use of the relationship Eq. (4.7) between ls-$g^*(s)$ and the concentration gradient, and, for clarity, use only the first term of the Fujita–MacCosham solution for the radial concentration profiles Eq. (2.31). It has gradients

$$\frac{d\chi(r,t)}{dr} = c_0 e^{-2\omega^2 st}\frac{m}{r}\frac{1}{2\sqrt{\pi Dt}}e^{-\left(\frac{m\left(\omega^2 st + \log(m) - \log(r)\right)}{2\sqrt{Dt}}\right)^2}, \tag{4.12}$$

which can be inserted in Eq. (4.7). Remapping the radial coordinate to s^* with Eq. (4.5) we obtain

$$g^*(s^*,t) = c_0\frac{m\omega^2 t}{\sqrt{\pi}\sqrt{4Dt}}e^{2\omega^2 t(s^*-s)}\times e^{-\frac{m^2\omega^4 t^2(s-s^*)^2}{4Dt}}, \tag{4.13}$$

where the leading term after the normalization factor is a Gaussian $e^{-(s-s^*)^2/2\sigma_s^2}$ [111] centered at s, with a variance

$$\sigma_s = \frac{\sqrt{2D}}{m\omega^2\sqrt{t}} \tag{4.14}$$

and fullwidth at half maximum

$$\Delta_s = \frac{4\sqrt{\ln 2}}{m\omega^2\sqrt{t}}\sqrt{D}. \tag{4.15}$$

This approximate analytical solution is depicted as blue crosses in Fig. 4.3 both in r-space and in s^*-space.

Not surprisingly, just like diffusion blurs the sedimentation boundaries in the original data space, it does the same in the transformed space of apparent sedimentation coefficients. Notably, Eq. (4.15) predicts that the Gaussian becomes sharper with time. An example is shown in Fig. 4.4. This may seem counter-intuitive at first, but is caused by the exponential time dependence of the transformation of r to s^*

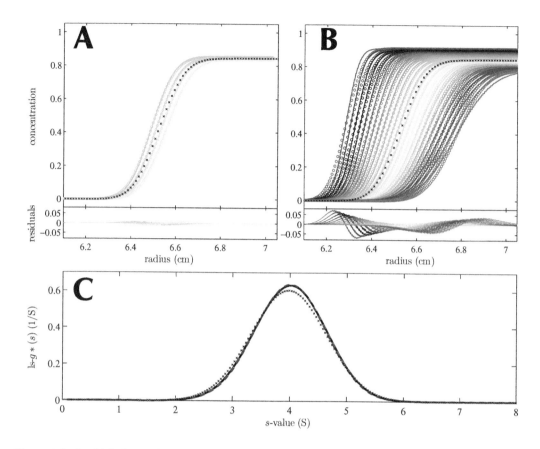

Figure 4.3 ls-$g^*(s)$ fit to ideal noise-free data of diffusing species (simulated for 80 kDa with 4 S, at 50,000 rpm in scan time intervals of 500 sec, circles show every 5th data point in each scan). Different time windows were included for the ls-$g^*(s)$ model (solid lines) with 250 steps from 0–8 S, not allowing for baseline offsets: *Panel A*: 7200–8100 sec (scans 24–27), leading to an rmsd of 0.0029; *Panel B*: 4200–11,100 sec (scans 14–37) leading to an rmsd of 0.0129. *Panel C*: Superposition of the corresponding ls-$g^*(s)$ distributions for A (black solid line) and B (magenta dashed line, virtually superimposing the black line). For comparison, shown are the concentration profiles predicted by the first term of the Fujita–MacCosham solution Eq. (2.31) for the same sedimentation parameters in the middle of the time window at $t_1 = 7650$ sec (crosses in A and B), and the corresponding analytical approximation of $g^*(s^*, t_1)$ following Eq. (4.13) (crosses in C). A fit of the Fujita–MacCosham solution Eq. (2.31) to the ls-$g^*(s)$ data between 2.5 and 5.5 S treating s and D adjustable parameters yields a virtually perfect fit (not shown) but leads to a ~10 % underestimate of diffusion with best-fit values of 4.03 S and 87.9 kDa.

(Eq. 4.5), which stretches the scale faster than diffusion takes place, such that the latter appears to broaden the distribution less with time. This is the equivalent of the point discussed above of the migration of sedimentation proceeding with $\sim e^{s\omega^2 t}$ and the migration from diffusion proceeding with a $\sqrt{t}$ scale; the same observation is now translated to the viewpoint of a frame of reference moving like sedimentation.

As outlined above, we cannot expect ls-$g^*(s)$ to fit well the sedimentation

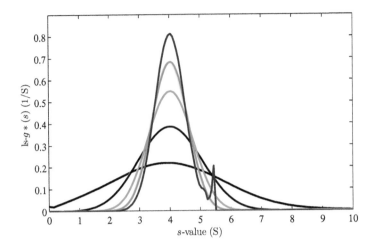

Figure 4.4 ls-$g^*(s)$ fit to the simulated data of a single diffusing species, as shown in Fig. 4.3, using different time points. Analyses shown are based on a 900-sec time interval centered at $t = 900$ sec (black), 3,000 sec (blue), 6,000 sec (cyan), 9,000 sec (green), and 12,000 sec (magenta). At the latest time, some back-diffusion was included in the analysis in order to demonstrate its effect on the distribution, which is that of a distortion and truncation at an upper s^*-limit. Due to the transformation of the radial scale, this upper limit for s^* appears at lower s-values for later times.

boundaries of a sedimenting species in the presence of significant diffusion, except for small time intervals.[10] This is illustrated in Fig. 4.3 for a 80 kDa, 4 S protein, which shows substantial residuals from the ls-$g^*(s)$ model when using a large time interval Panel B. However, it is a perhaps surprising (and theoretically not further examined) property of ls-$g^*(s)$ that it is remarkably well-behaved when applied to larger time intervals; the fits to the concentration profiles at different time points are balanced with symmetric residuals causing not much distortion of the resulting distribution as compared to the ideal distribution at the mid-point of the time interval. In the example of Fig. 4.3, both scan sets spanning short and long time intervals lead to virtually the same ls-$g^*(s)$ distribution. Limitations to this invariance will occur [152] for very broad boundaries of small particles when very large time intervals are used in conjunction with TI noise models. This is illustrated for a 40 kDa, 3.55 S protein in Fig. 4.5. Not surprisingly, for large time intervals the large residuals over extended broad boundaries will be compensated for by TI noise, which creates an artificial tail of ls-$g^*(s)$ towards lower s-values.

In conclusion, for a detailed quantitative analysis of the ls-$g^*(s)$ profiles, the time interval should be limited to a range that does not exhibit a substantial amount of

---

[10]The shortest possible time interval is zero, stemming from the analysis of only a single scan. Such a single-scan analysis can be done only in the absence of TI noise or other baseline offsets, since otherwise there is no basis for distinguishing the signal of the sedimenting particles from the baseline signal offsets. In the presence of TI noise, the signal/noise ratio for the part of the data unequivocally reporting on the sedimenting particles increases (i.e., the correlation of the signal of the sedimenting particles with the TI signal decreases) with increasing time interval of scans.

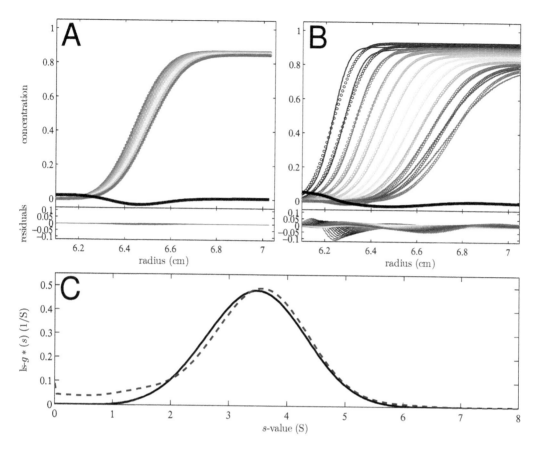

Figure 4.5 ls-$g^*(s)$ fit to data of a diffusing species, analogous to Fig. 4.3, but with broader boundaries and considering TI noise in the analysis. Boundaries are simulated for a 40 kDa, 3.55 S species at 50,000 rpm in scan time intervals of 30 sec, circles show every 5th data point in every 5th scan included in the fit). Time windows for ls-$g^*(s)$ fits (solid lines) were 7200–8400 sec in *Panel A*: (leading to an rmsd of 0.0018); and 3300–12,300 sec in *Panel B*: (leading to an rmsd of 0.0141). The TI noise from the fit is shown as a bold black line. *Panel C*: Superposition of the corresponding ls-$g^*(s)$ distributions for A (black solid line) and B (magenta dashed line).

systematic residuals (other than the checkered pattern caused by the use of step functions.) A greater tolerance for residuals is possible if TI and RI noise is not combined with the model, which is possible for the analysis of absorbance optical data. Generally, however, residuals in ls-$g^*(s)$ should be taken as an indication that diffusion is significant, i.e., that ls-$g^*(s)$ is not an appropriate model, and a model including diffusion should be applied, as discussed in Chapter 5.

Provided $g^*(s)$ is an accurate reflection of the sedimentation data for a diffusing species (i.e., ls-$g^*(s)$ leads to a good fit), then a second stage analysis is possible. By fitting a Gaussian Eqs. (4.13) and (4.14) an estimate of both s and D — and therefore via the Svedberg equation Eq. (1.35) also M_b — may be obtained. However, this is only of moderate precision (Fig. 4.3): With correct parameters the $g^*(s)$ distribution is quite different from the analytical prediction, or an excellent match

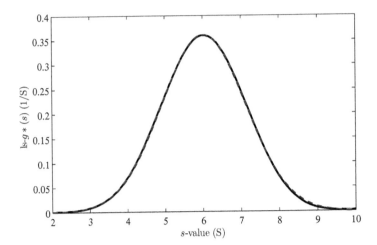

Figure 4.6 ls-$g^*(s)$ of the simulated sedimentation data of the heterogeneous mixture from Fig. 3.3B, consisting of a 100 kDa, 6 S species in superposition with an 80 kDa, 5 S species and a 120 kDa, 7 S species, both at 20% of the total signal (black solid line). Virtually indistinguishable (dashed magenta line) is the best Gaussian fit to ls-$g^*(s)$ using an impostor single-species model, resulting in an apparent s-value of 6.02 S and an apparent diffusion coefficient of 7.99 F, corresponding to an apparent molar mass of 68 kDa.

of $g^*(s)$ can be achieved with a significant underestimate of D (or overestimate of M). As shown by Philo, this can be addressed by transforming higher-order analytical approximations [153], or numerical Lamm equation solutions *via* the Bridgman relationship Eq. (4.7) [152] to generate a more accurate fitting model for $g^*(s)$. This can also account for artifacts due to large time intervals [152]. However, it cannot compensate for the loss of information associated with the large data misfit — including the opportunity to determine s_w *via* the transport method — that is unavoidable when using $g^*(s)$ to model diffusion processes over a significant time interval, or the lack of information when considering only a quasi-instantaneous data set. Most importantly, this eliminates the opportunity to exploit the time dependence of boundary broadening caused by diffusion *versus* differential migration of two or more species to learn about the nature of the sedimentation process.

This is illustrated in Fig. 4.6 for the three-component mixture previously shown to create significant misfit with an impostor single-species model when the entire time course of sedimentation is taken into account, but not in a fit of a small data subset (Fig. 3.3). Correspondingly, the ls-$g^*(s)$ profile of the data subset follows almost perfectly a single Gaussian with a vastly overestimated diffusion coefficient, and underestimated molar mass (Fig. 4.6).

CHAPTER 5

Distributions of Diffusing Particles

AFTER the excursion to the limiting case of apparent sedimentation coefficient distribution of non-diffusing particles, we focus back on trying to find a model that aims to embody the true physical process of sedimentation and the real properties of the sedimenting particles, and to stringently test the model by fitting to the complete set of available experimental data. Having laid the groundwork of defining and examining the characteristics of distributions and discussing the computation of distributions by direct least-squares fitting of the raw sedimentation data over the time course of the whole experiment, we can advance to the distributions of diffusing particles simply by exchanging the kernel of the integral equations from the step functions $\chi_{\mathrm{nd}}(s, r, t)$ of non-diffusing species to the Lamm equation solutions of real, diffusing species, $\chi(s, D, r, t)$.

5.1 TWO-DIMENSIONAL SIZE-AND-SHAPE DISTRIBUTIONS

A very general class of models that does not make assumptions about sedimenting particles other than their ideal sedimentation behavior is the family of two-dimensional size-and-shape distributions. They allow the description of species with any combination of hydrodynamic parameters sedimenting side-by-side in the sample mixture.

To simplify the notation, we will use in the following the symbol f_r for the frictional ratio instead f/f_0. There are several closely related size-and-shape distributions: $c(s, f_r)$, $c(s, M)$, $c(s, D)$, and $c(s, R_S)$, among other permutations of the inter-related hydrodynamic and thermodynamic parameters: For each sedimenting species, it holds that if any two parameters of the set $\{s, D \text{ or } R_S, f_r, M\}$ are fixed, the others are implicitly defined given $\bar{v}$ [154]. However, because the frictional ratio can only take values in a relatively small numerical range, and the sedimentation coefficient dimension is easily resolved, $c(s, f_r)$ is the variant most effectively discretized on a rectangular mesh and usually most efficiently calculated and displayed.

It is defined as [154]

$$a(r,t) \cong \int_{s_{min}}^{s_{max}} \int_{f_{r,min}}^{f_{r,max}} c(s,f_r)\chi_1\left(s, D(s,f_r), r, t\right) ds df_r + b(r) + \beta(t) \qquad (5.1)$$

with $\chi_1(s, D, r, t)$ denoting the Lamm equation solution at unit loading concentration. Since the Lamm equation depends on the diffusion coefficient and not on the frictional ratio, we make use of the functional dependence of D on s and f_r,

$$D(s,f_r) = \frac{\sqrt{2}}{18\pi} kT s^{-1/2} (\eta f_r)^{-3/2} \left(\frac{(1-\bar{v}\rho)}{\bar{v}}\right)^{1/2}, \qquad (5.2)$$

which follows from combining the Stokes–Einstein and Svedberg relationships Eqs. (1.34) and (1.35) [14]. Alternatively, for example, applying the Svedberg equation leads to $M(s,f_r)$, and in this scale the combined sedimentation coefficient/molar mass distribution is

$$a(r,t) \cong \int_{s_{min}}^{s_{max}} \int_{M_{min}}^{M_{max}} c(s,M)\chi_1\left(s, D(s,M), r, t\right) ds dM + b(r) + \beta(t), \qquad (5.3)$$

which is related to $c(s,f_r)$ with a scaling factor

$$c(s,M) = c(s,f_r) \times \left(\frac{dM(s,f_r)}{df_r}\right)^{-1}, \qquad (5.4)$$

i.e., the derivative of the coordinate transformation.

Figure 5.1 shows an example of the size-and-shape distribution in (s,M)-coordinates for interferometric sedimentation data of bovine serum albumin (BSA). It is apparent that, in contrast to $g^*(s)$, this distribution does not simply reflect the shape of the sedimentation boundary, but exhibits much more detail of the population of sedimenting species. For example, the BSA monomer and dimer peaks are well separated in the distribution, although they do not exhibit separate sedimentation boundaries. *As a consequence of accounting for diffusion in the kernel functions, computing the distribution essentially amounts to the deconvolution of diffusion from the sedimentation boundaries to expose the underlying distribution of sedimentation coefficients*, as well as their corresponding diffusion coefficients (or molar mass values or frictional ratios, respectively). Further, the quality of fit to the raw experimental data is excellent; this is a stringent and necessary condition for the accurate representation of the populations of species in the sample.

The dotted lines in Fig. 5.1B represent lines of constant f_r, which in the (s,M) coordinate system follows the well-known 3/2-power law Eq. (1.33). As indicated above, the reason why the choice of the (s,f_r) coordinate system is so attractive as a basis for the initial discretization is that the numerical values of f_r cannot be smaller than unity (provided the f_r-scale provided by the $\bar{v}$-value is close to the

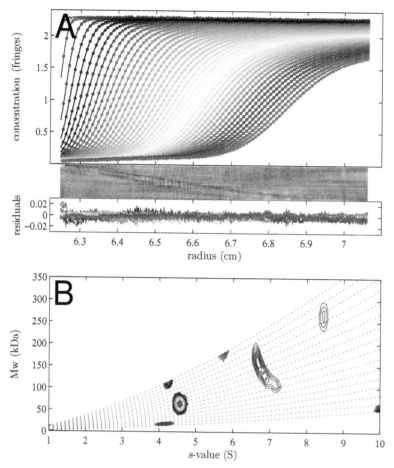

Figure 5.1 Size-and-shape distribution model $c(s, M)$ of a BSA sample sedimenting at 50,000 rpm [154]. *Panel A*: Experimental interference fringe shift data (circles, only every 10th data point shown), fit (solid line), and residuals overlay and residuals bitmap with an rmsd of 0.00389 fringes. *Panel B*: $c(s, M)$ distribution corresponding to the fit in A, presented as a contour map with color temperature indicating the amplitude of the distribution. Lines of constant frictional ratio are indicted as blue dotted lines. The structure and position of the minor peaks will depend on the details of the discretization.

physical scale) and is rarely larger than 2.0 for proteins or 3.0 for polymers. Therefore, an equidistant rectangular mesh in (s, f_r) coordinates can efficiently describe a distribution that, for example, in (s, M) coordinates would imply unphysical parameter combinations or very different particle densities.

Other than potential questions regarding the choice of a suitable f_r-range, it is important to note that for mixtures of species with dissimilar $\bar{v}$, for SV experiments in the absence of significant density gradients, one can operationally take a single $\bar{v}$-value for the computation of the distribution, and then later — if correct $\bar{v}$-values for species represented in a particular peak are available — correct the scale of M and f_r for this species. This involves no approximation in the calculation of the distribution and the sedimentation analysis, and just requires a transformation of the numerical values before further interpretation of these parameters. Alternatively, the distribution may be transformed to a $c(s, D)$ distribution, which is not dependent on $\bar{v}$. An example for the analysis of species with dissimilar partial-specific volumes can be found in [154], and the application to nanoparticles is described in [155].[1]

One small extension of Eq. (5.1) that is very useful in practice is to include a separate species to explicitly describe the interference signal contributions from the sedimentation of buffer salts unmatched in the reference sector,

$$a(r,t) \cong c_1\chi_1\left(s_1, D_1, r, t\right) + \int_{s_{\min}}^{s_{\max}} \int_{f_{r,\min}}^{f_{r,\max}} c(s, f_r)\chi_1\left(s, D(s, f_r), r, t\right) dsdf_r + b(r) + \beta(t)$$

(5.5)

accounted as a Lamm equation solution with s_1 and D_1 with the loading signal c_1. The reason is that buffer salts have a very different $\bar{v}$, such that the adequate description of the buffer signal would otherwise require covering of artificially low apparent f_r-values in the (s, f_r) mesh, just to cover this species. Using the extra discrete species allows us to maintain the lower physical limit of f_r (for given macromolecular density) as a constraint for the macromolecular species by virtue of constraining the (s, f_r) mesh.

The size-and-shape distribution model, with and without an extra discrete species, can be found in the Model menu in SEDFIT. All distributions are initially discretized with a s-f_r-mesh, which is efficient and intuitive after a preliminary $c(s)$ analysis providing the s-range and the average f_r. Different switches in the model parameter box can transform the distribution into different hydrodynamic or thermodynamic coordinate systems. It should be noted that no normalization occurs, i.e., integration is substituted by summation, which eliminates the need for analytic re-scaling factors. The default regularization is switched to Tikhonov–Phillips regularization, commensurate with the broader distributions to be expected in the f_r dimension. The final distribution is displayed in a superposition of a color heat map, as well as a general $c(s, *)$ (see below).

A closer inspection of Fig. 5.1 suggests that the size-and-shape distribution is limited in precision and/or resolution in the f_r (or M) dimension for the particles with low relative populations. This can be discerned, for example, from the elongated shape of the BSA dimer peak, which is broadened over a wide M-range by

---

[1]In the presence of significant density gradients, this approach is not feasible, since the SV experiment will exhibit an additional level of separation and $\bar{v}$ will modulate every species' sedimentation process, rather than just mapping M and f_r on different scales. For example, in isopycnic density centrifugation, the radial position of neutral buoyancy where each particle will ultimately band will be determined entirely by $\bar{v}$. For such cases, clearly the two-dimensional size-and-shape is insufficient, and one would need either a three-dimensional size-and-shape-and-density distribution, or a scale relationship determining $\bar{v}$ such as $\bar{v}(M, f_r)$.

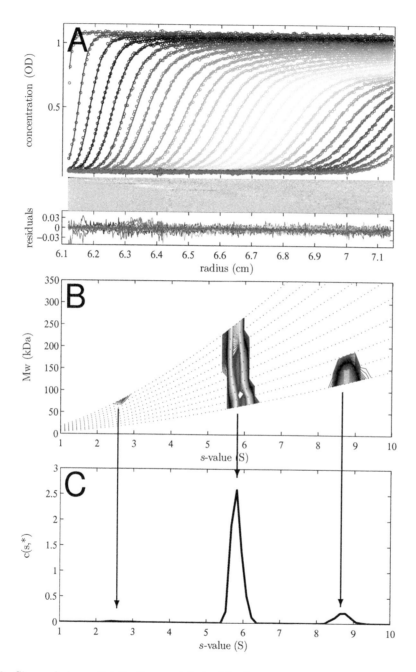

Figure 5.2 Size-and-shape distribution model $c(s, M)$ of an immunoglobulin G (IgG) sample sedimenting at 50,000 rpm [137]. *Panel A*: Experimental absorbance data (circles), fit (solid line), and residuals overlay and residuals bitmap with an rmsd of 0.0068 OD$_{280}$. *Panel B*: $c(s, M)$ distribution corresponding to the fit in A, presented as a contour map with color temperature indicating the amplitude of the distribution. Lines of constant frictional ratio are indicted as blue dotted lines. *Panel C*: General $c(s, *)$ obtained by integration of $c(s, f_r)$ in the f_r dimension.

regularization, indicating limited information content of the SV data on the diffusion of this species. In fact, variation of the discretization parameters can change the substructure and location of this peak (data not shown). This point is even clearer in Fig. 5.2, where even the main peak is not well resolved in the f_r (or M) dimension [137]. Contributing to the lower resolution are the lower number of scans, and the 4-fold lower signal-to-noise ratio.

There are three approaches to improve the analysis: condensing the result to the essential information, the inclusion of more data, and the implementation of constraints in the model. The first approach leads to the general $c(s, *)$, the second to extensions of global modeling, and the third to the $c(s)$ distributions. These will be discussed in sequence.

5.2   THE GENERAL $c(s, *)$ DISTRIBUTION

Among the most reliable pieces of information in an SV experiment are the sedimentation coefficients and their distributions. A straightforward way to extract this information from $c(s, f_r)$ is the integration over the f_r dimension, as shown in Panel C of Fig. 5.2. This takes the form

$$c(s, *) = \int_{f_{r,\min}}^{f_{r,\max}} c(s, f_r) df_r \tag{5.6}$$

and is referred to as the "general c-of-s" distribution [154].[2] The information that $c(s, *)$ provides is a pure sedimentation coefficient distribution, i.e., the representation of loading signal of particles with given s-values, irrespective of their shape or mass. No assumptions and approximations regarding the uniformity or the population of particles at a given s-value are made, nor how the f_r-values of populations with different s-values relate to each other. This indiscrimination in the second dimension is emphasized by the asterisk, in distinction from $c(s)$, which as been associated with sedimentation coefficients embedding scale relationships between the sedimentation and diffusion dimension (see below). In practice, though, $c(s, *)$ and $c(s)$ will be virtually identical in most cases.

> In SEDFIT, with fit of the size-and-shape distribution model, $c(s, *)$ is automatically calculated. It is shown as a black solid line in superposition to the $c(s, f_r)$ heat map, and can be retrieved for processing and plotting using the Copy Distribution function.

---

[2]Analogously, we can reduce the $c(s, M)$ distribution by integration over the M dimension, and likewise for other related size-and-shape distributions.

5.3   GLOBAL MODELING OF SIZE-AND-SHAPE DISTRIBUTIONS

Instead of adjusting the information content of the result, we can take the opposite approach of providing more informative experimental data. The idea of global modeling is to combine data from different experiments or different techniques [16], preferably orthogonal in information, to better define in combination what a single experiment analysis cannot well resolve. The approach of directly modeling the raw data of the experiments with a Fredholm integral equation describing the size-and-shape distributions is naturally suited to global modeling. In principle, combining different data sets only requires exchange of the kernels to represent the respective experiments, while the distributions are identical in the global model. A slight complication arises from the fact that it may not be possible to know with absolute certainty that the different experiments can be performed at exactly the same loading concentrations, and this necessitates multiplicative concentration scaling factors to be optimized alongside the distribution.

Experiments that can be expected to produce mutually complementary information are those that have a varying degree of sedimentation $vs.$ diffusion. Also they should take place in free solution and be "absolute" methods, excluding, for example, size exclusion chromatography due to the potential interactions with the matrix and the requirement for a calibration in terms of Stokes radii.[3] SV experiments at different rotor speeds are an obvious choice, since the variation of the rotor speed allows for the modulation of the sedimentation fluxes. Also, SE at different rotor speeds can contribute unique information, although a slight difficulty is the dependence of species' loading concentrations on the bottom position of the solution column, which is difficult to determine and should be treated as an additional floating parameter. Dynamic light scattering (DLS) can also be an excellent, orthogonal source of information contributing purely to the diffusion dimension. Here, the signal is dependent not only on the relative concentration of species, weighted by their molar mass, but also on the number of coherence areas that determine the amplitude of the autocorrelation function $g^{(1)}(\tau)$ [156].

An important additional consideration is that the optical signal utilized will reflect all species equally in each experiment. For example, the combination of absorbance and interference AUC data may introduce a mismatch. Similarly, DLS seems to naturally pair well with refractive index sensitive interference optics, but not with absorbance optical AUC data, unless weight-based absorbance coefficients (and separately the weight-based refractive index increments) of all species in the distribution are known to be identical. More complex considerations are required in the multi-method analysis for distributions of macromolecules with non-uniform $\bar{v}$.

Taken together, the following system of simultaneous equations describes the

---

[3]Obviously, for global analysis of data from different instruments, including different analytical ultracentrifuges, the calibration accuracy is particularly important. See Part I, Chapter 6 for AUC calibration methods.

model [157]

$$a_\omega^{(SV)}(r,t) \cong \gamma_\omega^{(SV)} \iint\limits_{s,D} c(s,D)\chi_1(s,D,r,t)dsdD + b_\omega^{(SV)}(r) + \beta_\omega^{(SV)}(t)$$

$$a_\omega^{(SE)}(r) \cong \gamma_\omega^{(SV)} \iint\limits_{s,D} c(s,D)c_{0,\omega}(\frac{sRT}{D},m,b)e^{\frac{s}{D}\omega^2\frac{1}{2}(r^2-m^2)}dsdD + b_\omega^{(SE)} \qquad (5.7)$$

$$g^{(1)}(\tau) \cong \gamma^{(DLS)} \iint\limits_{s,D} c(s,D)\frac{sRT}{D(1-\bar{v}\rho)}e^{-Dq^2\tau}dsdD \quad ,$$

where normalization factors γ depend on overall signal amplitudes (to be refined in the fit). The normalization factor $c_{0,\omega}$ in SE reflects equilibrium redistribution that is dependent on the rotor speed, buoyant molar mass sRT/D, and geometry of the solution column including meniscus and bottom [158]. Importantly, the distribution $c(s,D)$ is common to all models for all data sets. The discretized form of this model analogous to Eq. (3.5) is straightforward but tedious and will be omitted here.

One interesting problem encountered at this point when combining SV data with those of other techniques is how to choose proper weights for the different experiments in the least-squares optimization. A strict statistical choice of calculating weights by the number of data points and noise in the data acquisition would lead to a situation where SV data would completely outweigh all other experiments, simply by the very large number of data points (10^4–10^5). However, this does not capture the true value and information content of the respective experiments [16]. We should therefore deviate from this strict statistical view and remember that systematic errors in the different experiments are likely to be present in each experiment [16]. For example, SV is very much systematically dependent on the knowledge of the correct buffer viscosity and the absence of convection at the start of the run, factors that are not captured in the statistical error of data acquisition, and are completely irrelevant in SE. Further, SV poses much higher demands on the stability over time of the optical detection system, a factor not as critical in SE. Considering that each technique has unknown systematic errors, and the fact that from experience we obviously place trust in the information contributed from each technique, a logical choice for the relative weights is to adjust them in a way that allows each experiment to statistically contribute to the fit [16]. Empirically, this can be achieved by implementing relative statistical weight values that compensate for the number of data points [16].

To illustrate the gain in resolution of a size-and-shape distribution by introducing more, orthogonal experimental data, let us assume a model system of three proteins: a roughly spherical 40 kDa species with an f_r-value of 1.20, a 100 kDa protein species of the same roughly spherical shape, and a very elongated protein, also 100 kDa, with an f_r-value of 1.55 (perhaps a partially unfolded form of the larger protein). Each is present in quantities to produce the same signal.[4] If we only

---

[4]This is the same model system as used in the original introduction of two-dimensional size-and-

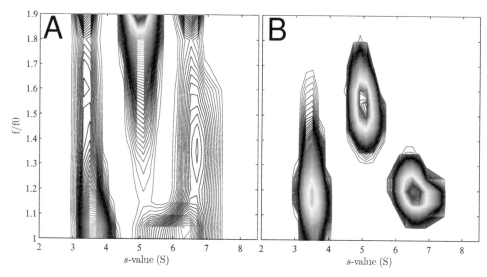

Figure 5.3   Effect of global multi-method modeling on the size-and-shape distribution model $c(s, f_r)$. *Panel A*: SV data were simulated at 55,000 rpm for a mixture of three macromolecular species with molar masses and frictional ratios of 40 kDa and 1.20, 100 kDa and 1.20, and 100 kDa and 1.55, respectively, at equal concentrations. Clearly, their diffusional properties are not well defined. *Panel B*: A global analysis of data for the same system including SV at 55,000 rpm and 20,000 rpm, DLS, and SE at 10,000 rpm, 15,000 rpm, and 20,000 rpm can resolve the mixture very well.

have SV data from a single experiment at 55,000 rpm, the resulting distribution is relatively well resolved in sedimentation coefficient, but poorly in the diffusion coefficient dimension (Fig. 5.3A). Including SV data at 20,000 rpm greatly improves the situation, since it allows for higher diffusion fluxes relative to the sedimentation fluxes (not shown). An even better resolution is achieved by additionally including DLS and SE data obtained at a range of rotor speeds (Fig. 5.3B).

Unfortunately, one significant drawback of this approach is that it is very time-consuming and demanding experimentally, such that it cannot be performed on a routine basis. Therefore, in the following we discuss a form of enriching the information of SV analysis by introducing prior knowledge of the properties of the sedimenting particles.

5.4   THE SEDIMENTATION COEFFICIENT DISTRIBUTION $c(s)$

Scale relationships for $D(s)$, $s(M)$, or equivalent, are the basis of one-dimensional sedimentation coefficient distributions $c(s)$ [14, 57, 159]. These allow species at each

shape distributions [154]. Whereas in [154] SV data at 50,000 rpm were considered, the data in Fig. 5.3A are at a slightly higher rotor speed of 55,000 rpm. This causes the sedimentation coefficients to be slightly better resolved, at the cost of the frictional ratios that are less well resolved. This somewhat more extreme situation demonstrates the impact of additional data containing diffusion information.

s-value to be unambiguously tied to a single f_r-value or molar mass. For many or most systems studied in practice, suitable scale relationships can be readily identified.

Before discussing the specifics of the different flavors of $c(s)$, we should summarize some properties of useful scale relationships that we can deduce already: (1) For each clearly visible sedimentation boundary, both the diffusional spread and heterogeneity in s can be unraveled simultaneously, provided a sufficiently long time range of data is available. Therefore, for each boundary that can be visually clearly discerned from the data, the scale relationships may have one adjustable parameter, such that theoretical expressions for the extent of boundary spread can fit to the experimental data.[5] This should allow for boundaries that would correspond to major, well-determined peaks in $c(s, M)$ to be faithfully represented also in $c(s)$. (2) For trace components we saw that a single, high-speed SV experiment may not have sufficient information to define the shape dimension very well. Conversely, we can anticipate that this will make it simpler to achieve a good fit of the SV data by the $c(s)$ distribution model, even if the scale relationship does not perfectly represent the diffusional spread of the species with low populations. However, the goal is to incorporate *a priori* knowledge about the system and rational expectations into the scale relationship, such that it will also provide the best possible estimate for D, M, and f_r of the minor components.

The general form of all variations of $c(s)$ is

$$a(r,t) \cong \int_{s_{\min}}^{s_{\max}} c(s)\chi_1(s, D(s), r, t)ds + b(r) + \beta(t) \tag{5.8}$$

with $D(s)$ determining the specific scaling law that describes the estimated extent of diffusion for species migrating with a certain s-value, as anticipated in Eq. (3.1). If $D(s) = 0$ for all s-values, this is identical to the ls-$g^*(s)$ model and the resulting distribution will be limited in resolution essentially to what one can discern from the visual inspection of the experimental SV data (after removing TI and RI noise). This is different for $c(s)$ with $D(s) > 0$: here, we can expect the distribution to be deconvoluted from the effects of diffusion and to have increased resolution. With the right choice of $D(s)$ appropriate for the sample under study, and with data of good signal/noise ratio, $c(s)$ will be a good approximation of the true sedimentation coefficient distribution.

5.4.1 Scaling Law Based on an Average Frictional Ratio $f_{r,w}$

When working with folded proteins, we can exploit the fact that their frictional ratio values fall usually within a fairly limited region of values between 1.2 and

---

[5]From our experience, more than one parameter describing the boundary spread per boundary are not well-determined — this can also be deduced from the fact that one is sufficient to provide a virtually perfect fit for most cases.

2.0, with some exceptions. At the root of this is the relative compactness of folded macromolecules, and the degeneracy of the relationship between a spatial structure that is three-dimensional and the translational friction coefficient that is only a single number. The back-of-the-envelope estimate that $s \sim M^{2/3}$ (Part I, Chapter 1), which has been used traditionally in the SV literature in the absence of other information, is based on the same assumption.

The scale relationship corresponding to this idea arises from the general relationship $D(s, f_r)$ in Eq. (5.2). The restriction to a single, average frictional ratio $f_{r,w}$ leads to

$$D(s) = \frac{\sqrt{2}}{18\pi} kT (\eta f_{r,w})^{-3/2} \left(\frac{(1 - \bar{v}\rho)}{\bar{v}} \right)^{1/2} s^{-1/2}, \tag{5.9}$$

where $f_{r,w}$ is used as an adjustable parameter for modeling experimental SV data [14,57].[6] Although there has been no rigorous proof for this, $f_{r,w}$ can be expected to be close to the signal-average f_r-value of all sedimenting species [57]. Thus, if there is one major species present in the sample, the best-fit $f_{r,w}$-value will be a very good approximation of the true f_r-value of that species. It is emphasized that for Eq. (5.9) to be exact, different particle populations are not necessarily required to have the same shape, because many different shapes can result in the same f_r-value.

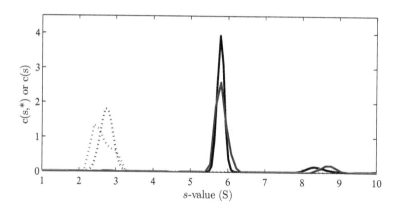

Figure 5.4 Comparison between the $c(s, *)$ distribution (black) from Fig. 5.2 and the $c(s)$ distribution (magenta) applied to the same data. The fit quality from $c(s)$ with a single best-fit $f_{r,w}$-value is very similar, with a slightly higher rmsd of 0.00695 OD$_{280}$. The dotted lines are the same distributions shown on a 100-fold expanded scale between 1 and 5 S, to highlight the detection of trace species.

An example for the relationship between $c(s, f_r)$, $c(s, *)$, and $c(s)$ is provided in Figs. 5.2 and 5.4. The $c(s)$ model with single a best-fit average single friction ratio fits the data almost equally well as the $c(s, f_r)$ model. The s-values of the major species are essentially identical, and the minor species are very close. Whereas the

---

[6]It is easy to confirm that with $D \sim s^{-1/2}$ and $M \sim s/D$, the scaling law Eq. (5.9) is consistent with the traditional $s \sim M^{2/3}$ law.

apparent molar mass values were poorly defined in the two-dimensional distribution, the best-fit $f_{r,w}$ of 1.59–1.67 (one standard deviation confidence interval) implies apparent molar mass values (uncorrected for solvent density and viscosity) of ~131–140 kDa for the ~6 S species and ~237 kDa for the ~8–9 S species, revealing the latter to be the dimer. The very sparsely populated ~3 S species is implied to be a ~44 kDa species (within error consistent with IgG fragments).

It is interesting to examine more closely how well $f_{r,w}$ is determined by the experimental data, and how the distribution shape varies with $f_{r,w}$. To this end, Fig. 5.5 compares $c(s)$ distributions obtained with different, sub-optimal values for $f_{r,w}$. First, it can be discerned that there is a well-defined minimum in the error surface for $f_{r,w}$. In the limiting case of $f_{r,w}=\infty$, which is equivalent to the ls-$g^*(s)$ model, the quality of fit is very poor as judged from both the rmsd and the systematic residuals bitmap pattern, and the corresponding $c(s)$ peak is very broad. As the $f_{r,w}$-value gets smaller and approaches the best-fit value of 1.58, the rmsd of the fit drops sharply and the systematic pattern in the residuals diminishes strongly. At the same time, the corresponding $c(s)$ peaks become sharper and start to baseline-resolve the dimer species at 9.3 S as a result of increasing deconvolution of diffusion. At the optimal value of 1.58, a near-perfect fit is achieved. Interestingly, at $f_{r,w}$-values smaller than the best-fit value, the rmsd strongly increases again and at the theoretical lower limit of 1.0, the fit becomes similarly poor as for the case of no diffusional deconvolution $f_{r,w}=\infty$. The position of the main peak is virtually unchanged for all $f_{r,w}$-values. However, because the dimer sediments close to the leading edge of the diffusionally spread sedimentation boundary of the monomer, the trace dimer peak does vary slightly with $f_{r,w}$ [59]. Such changes are smaller or absent for species sedimenting further apart from the monomer boundary.

The features shown in this example are typical for $c(s)$ analyses. The reason for the existence of a well-defined best-fit value for $f_{r,w}$ lies in the ability of SV data collected over a sufficient time interval to provide information on the diffusional spread independent of heterogeneity, as discussed above. Confidence limits will depend on the particular system studied, the boundary structure, the signal/noise ratio of the SV detection, as well as the time range of scans available for analysis.

Examples for the potential of $c(s)$ to hydrodynamically resolve similar-sized species are shown in Fig. 5.6 for medium-sized proteins, and in Fig. 5.7 for small macromolecules. In our earlier model system Figs. 3.2C and 3.3 on p.57/59 with a mixture of species with 80 kDa, 5 S (20%), 100 kDa, 6 S (60%), and 120 kDa, 7 S (20%), the species cannot be resolved on a signal/noise level of 100:1, and only one broad peak is obtained in a fit with excellent rmsd. (However, excellent molar mass estimates are obtained, as outlined in greater detail below; see Fig. 5.11.)

In this regard, it is important to realize that regularization is essential for the $c(s)$ distribution, as it is for all other distributions. If the same formalism is used as outlined so far for $c(s)$ but without regularization, we refer to this distribution as "pseudo-$c(s)$." As shown in Figs. 5.6 and 5.7, the pseudo-$c(s)$ results in sharp, baseline-separated peaks that provide the illusion of a much higher resolution and much more detail than is possible to reliably extract. This is an effect of noise

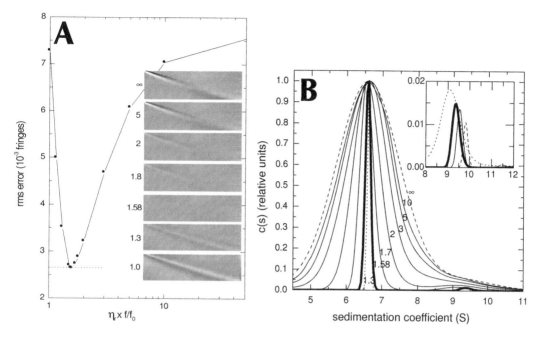

Figure 5.5 Relationship between $f_{r,w}$, fit quality, and distribution shape in $c(s)$. Interference optical sedimentation profiles obtained from an SV experiment similar to Figs. 5.2 and 5.4 (taken from [57]) are fit to a $c(s)$ model using different values of $f_{r,w}$. *Panel A*: The resulting fit quality, with rmsd values (solid circles), and residual bitmaps for representative values. *Panel B*: Some of the corresponding $c(s)$ distributions, with the inset highlighting the region of the dimer. The $c(s)$ distribution from the overall best fit with $f_{r,w}=1.58$ is shown in bold, the overly deconvoluted case with $f_{r,w}=1.3$ is shown with a dotted line, and the limiting case of $f_{r,w}=\infty$ (equivalent to ls-$g^*(s)$) is shown as a dashed line. In the inset, the distributions shown are only for $f_{r,w}$-values of 1.7 (dashed line), 1.58 (bold solid line), 1.55 (solid line), and 1.30 (dotted line). With the number of data points of ∼300,000, the curves for 1.58 and 1.55 are statistically indistinguishable on a confidence level of 2 standard deviations, suggesting an uncertainty in the dimer peak position of ∼0.3 S. The different heights of the peaks in the inset are a result of the normalization of the main peak.

amplification characteristic for the naïve solution of integral equations. In fact, re-analysis of the same sedimentation process with slightly different discretization parameters, or another draw of random noise in the data (of the same standard deviation) results in a different pattern of peaks (Fig. 5.6). This phenomenon was already anticipated and discussed in the general context of distributions in Section 3.5, along with tools for determining confidence limits (see Fig. 3.8 on p. 74).

Observation over a long time course covering the entire sedimentation process is essential for reliable discrimination of sample heterogeneity and diffusion in $c(s)$. As illustrated in Fig. 5.7 this is true in particular for small species with broad, diffusion-dominated boundaries. If only half of the data in Fig. 5.7 are considered, a correlation between $c(s)$, $f_{r,w}$, and the baseline occurs. With the baseline being

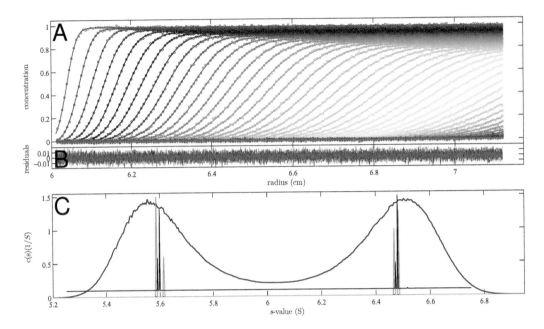

Figure 5.6 Potential for hydrodynamic resolution of medium-sized species with $c(s)$. *Panel A*: Shown are simulated sedimentation data for an mixture with equal signals from a 90 kDa, 5.6 S species, and a 115 kDa, 6.5 S species sedimenting at 50,000 rpm (in 300-sec intervals with noise of 0.005, every 5th data point shown). Residuals of a $c(s)$ fit are shown in *Panel B*. *Panel C*: The best-fit $c(s)$ distribution with regularization on a confidence level of one standard deviation (magenta), and without regularization in the same (green) and slightly different discretization (blue, offset by 0.1). The distributions without regularization are scaled by a factor 0.01.

poorly defined, the $c(s)$ peaks occur at slightly too low s-values (dashed lines in Fig. 5.7C).

The recognition of three peaks in Fig. 5.7 clearly shows the resolution power of $c(s)$ to deconvolute strongly diffusion-broadened boundaries. However, since only a single, average frictional coefficient is used, the accuracy in apparent molar mass values is limited, here resulting in relative errors of -1.6%, -14.5%, and -23.1%, respectively. Generally, the best-fit $f_{r,w}$ leads to reliable apparent molar mass estimates only for the case of a single major $c(s)$ peak that gives rise to the sedimentation boundary; otherwise $f_{r,w}$ degenerates into an adaptive deconvolution parameter that merely allows us to visualize the underlying sedimentation coefficient distribution. On the other hand, for our earlier model system Figs. 3.2C and 3.3 on p.57/59 the associated best-fit frictional ratio is 1.28, corresponding to an average molar mass close to the correct average (see Fig. 5.11). Thus, while not always entirely successful, the performance of the $c(s)$ model for molar mass determination is significantly better than models not accounting for polydispersity, and often the best available option.

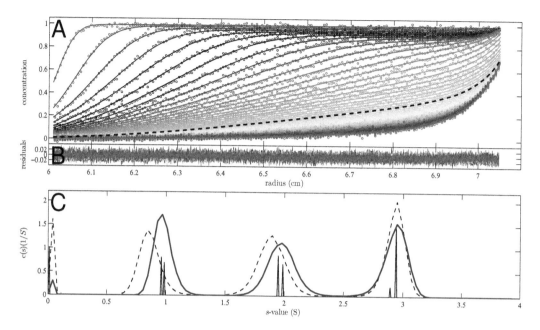

Figure 5.7   Example for hydrodynamic resolution of small species with $c(s)$. *Panel A*: Shown are simulated sedimentation data for an mixture with equal signals from a 6 kDa, 1 S species (f_r=1.16), a 20 kDa, 2 S species (f_r=1.30), and a 40 kDa, 3 S species (f_r=1.37) sedimenting at 50,000 rpm (in intervals of 600 sec, with noise of 0.010, every 5th data point shown). Residuals of a $c(s)$ model with best-fit $f_{r,w}$=1.20 are shown in *Panel B*. *Panel C*: The best-fit $c(s)$ distribution with regularization on a confidence level of one standard deviation (magenta), and without regularization (blue, scaled by factor 0.01). When only half the time course is analyzed— up to the dashed black line in A— the best-fit $f_{r,w}$ is 1.23, with a $c(s)$ distribution shown as a dashed black line in C.

The standard $c(s)$ model can be selected by clicking on `Continuous c(s) Distribution` in the `Model` menu in `SEDFIT`. After fitting the model, the `Display` function `Show Peak Mw in c(s)` (alternatively invoked using the Ctrl-M keyboard shortcut) produces buttons near each peak labeled with the apparent molar mass implied by the s-value and $f_{r,w}$.

When pressed, a message box appears that reports the integrated equivalent loading concentration, the s_w-value from the peak, a correction to standard condition, the Stokes radius, and axial ratios of the hydrodynamic equivalent ellipsoids.

In addition, this function will indicate the migration of this species in the boundary data by retracing the boundaries in red/grey color scheme, with the red intensity being approximately proportional to the calculated concentration gradient of an ideal species with the given sedimentation parameters. This allows us to visualize directly which $c(s)$ peak corresponds to which sedimentation boundary in the raw data space, the relative contribution of different species to the signal within one boundary, and the diffusional spread of each species.

5.4.2 Segmented $c(s)$ Distributions with Multiple Frictional Ratios and Hybrid Discrete/Continuous Distributions

In many applications, it is advantageous to use a variation of $c(s)$ in which an extra species is modeled separately, analogous to Eq. (4.2), in order to account for signal offsets arising from buffer salts or other small molecules. It is advantageous not to include the latter in the description via the weighted average frictional ratio $f_{r,w}$, since their partial-specific volume is frequently very different from that of the macromolecules of interest, such that their apparent f_r on the $\bar{v}$-scale of the macromolecule would introduce a distortion into $f_{r,w}$. Because of the shallow gradients from sedimenting small molecules, there is usually little cross-correlation of their parameters to the description of the faster-sedimenting particles of interest, if they are accounted for separately, even though the diffusion and back-diffusion of the small species can out-run sedimentation of the macromolecules. This is depicted in Fig. 5.8A.

Similarly, SV data often show visually distinct boundaries, as in Fig. 5.8B, which allows the independent refinement of diffusional scale relationships for each boundary component. In this case, multiple $f_{r,w}$-values may be determined on different branches of the $c(s)$ distribution with little cross-correlation.

In general, the distribution can be constructed as a sum of multiple discrete species in combination with several continuous segments

$$a(r,t) \cong \sum_{i=1}^{I} C_i \chi_1(s_i, D_i, r, t) + \sum_{j=1}^{J} \int_{s_{\min}}^{s_{\max}} c(s)\chi_1(s, D^{(j)}(s), r, t)ds + b(r) + \beta(t). \quad (5.10)$$

This is termed hybrid discrete/continuous distribution with I discrete species and J continuous segments, where each of the $c(s)$ segments describes a different, non-overlapping s-range corresponding to a separate boundary, and each segment may be based on a separate scale relationship $D^{(j)}$ resting on different scale parameters $f_{r,w}^{(j)}$.[7]

An example for the application of this approach is shown in Fig. 5.9, which provides an analysis of the same data of BSA as shown in Fig. 5.1 (p. 93). However,

---

[7]The difficulty arises as to how to display the assembly of discrete species and continuous segments. The default in **SEDPHAT** is the convention that the discrete species be plotted with the vertical axis interpreted directly as loading concentrations, whereas the continuous segments continue to be "normalized," i.e., the integral over the distribution represents loading concentrations. Alternatively, avoiding the double meaning of the vertical axis, one can modify the continuous segments omitting the multiplication with the discretization width Δs_i in Eq. (3.5), which then displays all species that define the grid of s-values directly in terms of their loading concentrations. A disadvantage of the latter approach is that the contributions from the grid points in the continuous segments will depend on their respective grid spacing. *Vice versa*, one could in principle assign the discrete species an artificial grid size (e.g., half the distance between the edges of the surrounding continuous segments), in which case all the elements of the hybrid discrete/continuous distribution would be in terms of differential distributions, i.e., integration would lead to the loading concentrations.

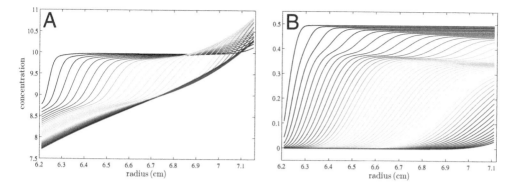

Figure 5.8 Examples for distinct boundaries justifying the application of a segmented $c(s)$ model. *Panel A*: A 150 kDa, 7 S species in superposition with a small molecule of 0.2 kDa, 135 F (e.g., a buffer component) sedimenting at 50,000 rpm recorded in 300-sec intervals. *Panel B*: A 50 kDa, 4 S species in superposition with a 350 kDa, 10 S species.

instead of a constraint-free size-and-shape distribution $c(s, M)$, we now utilize our knowledge of the sample identity: We describe the main BSA monomer boundary as a discrete species with unknown s_1 and M_1, and the peaks from small oligomers are discrete species with unknown s-values but M-values known to be integral multiples of M_1. This leads to a well-determined apparent molar mass of the monomer, as well as s-values of the oligomers. However, the underlying assumption must be kept in mind that the peaks represent discrete species, which would be invalid in the presence of micro-heterogeneity (e.g., from heterogeneous glycosylation of proteins).[8,9]

The hybrid discrete/continuous model also offers the opportunity to broadly describe additional and contaminating species that sediment at sufficiently different rates than the species of interest (Fig. 5.9), which is a distinct advantage over a model consisting solely of a few discrete Lamm equation solutions. Often, the frictional ratio of $c(s)$ segments that describe only trace populations of sedimenting species is not well defined. However, this may be irrelevant if the s-range of these $c(s)$ segments is well separated from species of interest. In this case, the $f_{r,w}$-values may be fixed at hydrodynamically reasonable values.

Just like the size-and-shape distributions, the hybrid discrete/continuous models can be applied globally to multiple data sets, rendering Eq. (5.10) into a set of simultaneous equations, which, similar to Eq. (5.7), may span different techniques. Carrying out different experiments may easily introduce variations in sample

---

[8]If this assumption was incorrect, as always, representation as a discrete species would cause an overestimate of diffusion and a corresponding underestimate of the apparent molar mass, respectively, as discussed above.

[9]Similarly, it can be useful to constrain the ratio of s-values of different discrete species, for example, when implementing hydrodynamic models of oligomeric species. An example can be found in the analysis of tubulin rings [55].

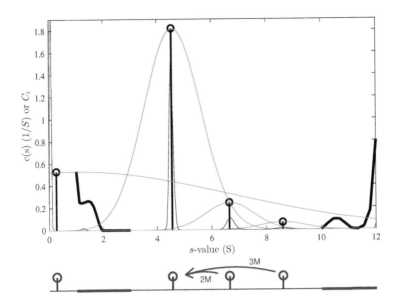

Figure 5.9 Analysis of the sedimentation data of BSA from Fig. 5.1 using the hybrid discrete/continuous distribution. The schematics on the bottom in red highlights the design of the distribution model. Major signal boundaries from monomeric and oligomeric BSA are modeled as discrete species (circles), with dimer and trimer constrained to have 2- and 3-fold the molar mass of the monomer, respectively. Additionally, a small discrete species represents sedimenting buffer components (left circle). This is combined with two continuous $c(s)$ segments modeling breakdown products and impurities smaller than the monomer, and higher oligomers and aggregates, respectively (bold blue lines, magnified 10-fold). The hybrid discrete/continuous model was constructed after initially carrying out a standard $c(s)$ analysis (magenta) providing estimated s-values for the monomer, dimer, and trimer, along with an estimate for the apparent monomer mass. The separation of the continuous segments from discrete species was chosen so as to avoid overlap and correlation. To this end, the approximate diffusional broadening of the boundaries of the discrete species is displayed in units of apparent s^*-values following Eq. (4.13) (green lines). The discrete s- and M-values are refined in non-linear regression, maintaining the mass ratio of 2:1 and 3:1 for dimer and trimer to monomer.

concentrations from small pipetting errors, which can be accounted for by allowing for concentration normalization factors for all segments and different concentrations for all discrete species. Further, for cases where the continuous segments describe only minor species that may vary from experiment to experiment, such as aggregation or other degradation products, the continuous $c(s)$ segments may be declared "local." In this case, separate $c(s)$ segments are calculated for each experiment.

The hybrid discrete/continuous distribution is available in SEDPHAT in variations that differ in the behavior of the model in global analysis: In the Hybrid Local Continuous Distribution and Global Discrete Species model the sedimentation parameters of the discrete species are global parameters (though not their concentrations), whereas the Hybrid Global Continuous Distribution and Global Discrete Species treats all as global parameters (with a scaling parameter for each experiment). The best choice depends on whether the $c(s)$ segments describe species of interest, or, e.g., impurities that may differ in different data sets.

As an aid to verify that the different discrete species are localized in separate sedimentation boundaries in these models, the distribution window displays a button labeled disc. g*(s,tmax), which when pressed draws the apparent sedimentation coefficient distributions Eq. (4.13) (similar to the green lines in Fig. 5.9).

In SEDFIT, three variations of the standard $c(s)$ model offer some functionality of the hybrid discrete/continuous model: In the Continuous c(s) with other prior knowledge submenu, the function Continuous c(s) with 1 discrete component allows the combination of a single discrete species with a single $c(s)$ branch. This slight extension of $c(s)$ is highly useful, for example, to include buffer signals in the model while excluding them from $f_{r,w}$ of $c(s)$. The function Continuous c(s) with bimodal ff0 offers two branches of $c(s)$, appropriate for two clearly separating boundaries to be fitted. Finally, the function Continuous c(s) with bimodal ff0 and 1 discrete component combines both features.

5.4.3 $c(s)$ with Constant Diffusion Coefficient or with Known Frictional Ratio

Another possibility to include hydrodynamic prior knowledge into the $c(s)$ model exists when studying polydisperse sets of molecules with the same diffusion coefficient, arising from an identical contour but different mass (and therefore different sedimentation coefficient). This applies, for example, to hollow structures that are partially filled, such as viral capsids with different degrees of nucleic acid load, or ferritin shells with a variable number of iron atoms inside. Since the diffusion coefficient depends only on the contour of the particle, not on the mass, the scale relationship $D(s)$ degenerates to $D(s) = D_0$, with the diffusion coefficient D_0 being uniform throughout the particles of the distribution. Interestingly, for such particles the resolution in sedimentation coefficients is proportional to the molar mass, $s \sim M$, which is higher than the usual $s \sim M^{2/3}$. D_0 can usually be treated as an adjustable parameter refined in the non-linear regression. An example for the application of this model to ferritin can be found in Figure 4 of [14].

On the $\bar{v}$-scale of the protein, hollow shells of identical outer contour and different total mass (from different void volumes inside, and/or different loading) have different frictional ratios. This is due to the mass dependence of the radius R_0 of the equivalent compact, smooth sphere with given density $1/\bar{v}$. On the other hand, since analytical ultracentrifugation will not reveal internal structure of particles, one could apply an equivalent picture where the particle is considered to be internally uniform, and of average density (compare the "play dough" and "sedimenting particle" frameworks introduced in Part I, Chapter 2). In this framework, particles of different mass and identical contour have identical frictional ratios, but different (particle-averaged) densities. In fact, we may rearrange Eq. (5.9) and determine the equivalent $\bar{v}_{SP}$ for a given s-value

$$\bar{v}_{SP}(s) = \left[\rho + sD^2 (\eta f_{r,w})^3 \left(\frac{18\pi}{2kT} \right)^2 \right]^{-1} \qquad (5.11)$$

provided we also know the particle frictional ratio.

A different flavor of $c(s)$ is based on a known frictional ratio $f_{r,w}$, for example, determined by electron microscopy. In this case, $f_{r,w}$ is fixed in the analysis and the relationship $D(s)$ of Eq. (5.9) is used with $\bar{v}_{SP}$ as an adjustable parameter, refined such that the observed degree of diffusion in the experimentally observed boundary at a certain s-value is well described. In contrast to the $c(s)$ model with constant diffusion coefficient where all species had the same contour, here the diffusion coefficients do change with particle size but the ensemble of particles has uniform density. Such a determination of the particle $\bar{v}_{SP}$ can be highly useful, for example, in the study of detergent-solubilized membrane proteins, where $\bar{v}_{SP}$ provides information on the protein/detergent ratio [160].

The $c(s)$ model with constant D is invoked in SEDFIT with the function Continuous c(s) with invariant D in the Continuous c(s) with other prior knowledge submenu. The $c(s)$ model with fixed $f_{r,w}$ is located in the same submenu. For homogeneous samples — in the absence of polydispersity — all variations of $c(s)$ will lead to the same result, but the models still exhibit different practical functionality because they allow refinement of different parameters, which eliminates operations to transform parameters between different scales. In the present case, either D or $\bar{v}_{SP}$ can be refined in the non-linear regression.

5.4.4 $c(s)$ in Density Contrast Experiments

From the preceding section it is apparent how $\bar{v}_{SP}$ and $f_{r,w}$ are interrelated. If data are available reporting on the sedimentation of the identical sample in solvents of different density, both $\bar{v}_{SP}$ and $f_{r,w}$ may be determined simultaneously in a global analysis [65]. There are limitations in the range of particle and solvent densities for which this analysis is possible, as described in detail in Part I, Chapter 3.

Similar to the global modeling for size-and-shape distribution Eq. (5.7), scaling factors γ_{xp} are necessary in the global density contrast analysis to account for unavoidable uncertainties in the precise sample concentration after dilution into different solvents. Extending the concept of segmented distributions Eq. (5.10), different $\bar{v}_{SP}$ values can be assigned to different $c(s)$ segments. This can be useful, for example, to allow the determination of $\bar{v}_{SP}$ of certain fractions of interest in samples of imperfect purity [65].[10]

---

[10]A variation of this theme, calculating a full distribution of particle densities as $c(s, \bar{v})$ at constant f_r has been described by Demeler [161]. From Eq. (5.11) it can be seen that this is essentially

Global analysis of SV data sets acquired at different densities is possible only in SEDPHAT, where the Hybrid Global Continuous Distribution and Global Discrete Species model can be combined with adjustable $\bar{v}$, if the Options function Fit VBARO, Invariant Particle Model is invoked. The latter appears in the menu only after AUC data with different experimental densities are loaded. In the implementation in SEDPHAT, an unknown $\bar{v}_{SP}$ can be ascribed to species in a range of s-values determined by the user, whereas a different (known) $\bar{v}$ is assigned all species outside this range.

5.4.5 $c(s)$ with Segments of Constant Molar Mass: Conformational Changes

This model is constructed such that species in a certain s-range have the same molar mass. This is achieved by excluding this region from use of the standard $f_{r,w}$-based scale relationship $D(s)$, and instead using the Svedberg equation Eq. (1.35), in a design illustrated in Fig. 5.10. This is appropriate, for example, when studying proteins of given molar mass that might be in an ensemble of different long-lived conformational states, leading to different sedimentation coefficients.[11] The molar mass can be used as an adjustable parameter to be determined from fitting the experimental data.[12]

This adaptation of the standard $c(s)$ has been applied to the study of ligand-induced conformational states of the rotavirus protein NSP2 [20]. This protein forms an octamer of 280 kDa that can undergo a conformational change from a 12 S species to a 12.6 S species upon binding of a nucleotide, and assumes a third state with long-lived tetramers and octamers at 11.8 S after binding Mg++.

Figure 5.10 Schematic of the construction of a $c(s)$ model for proteins with conformational change. A specific s-range (magenta) is excluded from the standard $c(s)$ (red). While the standard $c(s)$ segments use the hydrodynamic scale relationship based on an $f_{r,w}$ (shared for both segments), in the excluded region $D(s)$ is based on the Svedberg equation for a given constant molar mass.

---

identical to an earlier size-and-shape distribution $c(s, f_r)$ [154] *modulo* a transformation via (5.11) from the f_r to the $\bar{v}$ coordinate, as previously observed by Carney and colleagues [155].

[11]Conformational states that rapidly interchange on the time scale of sedimentation will only exhibit a single, time-averaged sedimentation coefficient (see Part I, Section 2.2).

[12]It is preferable that this be done even when the molar mass is known from other sources, due to the unavoidable errors in the partial-specific volume. Use of a predetermined molar mass value is recommended only when it has been experimentally determined in the AUC and is expressed on the same $\bar{v}$-scale.

5.4.6 Distributions of Wormlike Chains

Different hydrodynamic scale relationships exist for macromolecules that do not assume specific globular structures and instead consist of unstructured chains. The wormlike chain model may useful, for example, when studying polydisperse mixtures of some linear polymers, completely unstructured proteins, nucleic acids, or size distributions of flexible fibrils [162].

The model of wormlike chains can be imagined as a long cylinder of a certain diameter that can bend [163]. For a molecule of given mass and density, the mass per length allows us to determine the chain diameter and length. A key aspect of wormlike chains is their flexibility, which is characterized by the persistence length, i.e., the distance along the chain where the correlation between the angle of tangential vectors along the chain has decayed to $1/e$. Hydrodynamic friction coefficients for wormlike chains are smaller than those of stiff rods, since the chain can to some extent bend back onto itself and diminish resistance. Analytical expressions for configurational averaged translational friction coefficients of wormlike chains as a function of contour length L, diameter d, and persistence length q have been derived by Yamakawa and Fujii [164]. For example, for chains longer than $4.556q$ it is[13]

$$f = \frac{3\pi\eta L}{A_1(L/2q)^{\frac{1}{2}} + A_2 + A_3(L/2q)^{\frac{-1}{2}} + A_4(L/2q) + A_5(L/2q)^{\frac{-3}{2}}} \quad (5.12a)$$

with

$$A_1 = (4/3) \times (6/\pi)^{1/2}$$
$$A_2 = -\left(1 - 0.01412d^2 + 0.00592d^4\right)\log d$$
$$\quad - 1.0561 - 0.1667d - 0.1900d^2 - 0.0224d^3 + 0.0190d^4$$
$$A_3 = 0.1382 + 0.6910d^2 \quad (5.12b)$$
$$A_4 = -(0.04167d^2 + 0.00567d^4)\log d - 0.3301 + 0.5d - 0.5854d^2$$
$$\quad - 0.0094d^3 - 0.0421d^4$$
$$A_5 = -0.0300 + 0.1209d^2 + 0.0259d^4 \quad .$$

Combined with the Svedberg relationship, Eq. (5.12b) allows us to solve for the diffusion coefficient associated with a species of given sedimentation coefficient $D(s)$. Alternatively, Monte-Carlo approaches for random chains have been described by Zimm [166] and later by MacRaild *et al.* [162].

For some macromolecules forming wormlike chains, the chain diameter is not immediately obvious, but the mass per length m_L may be determined experimentally, for example, by scanning transmission electron microscopy [167], or in atomic force microscopy [168]. This allows the cross-section area to be calculated as $A_c = m_L \bar{v}$, and the diameter follows as $d = 2\sqrt{A_c/\pi}$.[14] If the sedimentation data contain

---

[13]With the corrections indicated in [165], but not considering the extra term in A_4 of [165].

[14]If the mass per length is anhydrous, then the usual inflation to estimate hydrodynamic contours should be applied.

diffusion information, then the persistence length can be treated as an adjustable parameter and refined in the fit, replacing $f_{r,w}$ as a parameter extracting the diffusion information.

> The $c(s)$ variant for wormlike chains in SEDFIT is invoked in the Continuous c(s) with other prior knowledge submenu with the function Continuous c(s) for wormlike chains. Motivated originally by the analysis of fibrils, the mass per length and chain diameter is determined by the fibril anhydrous cross-section in combination with a hydration parameter. For chains shorter than one tenth of the persistence length, the hydrodynamic friction is taken as that of a linear rod. A monomer molar mass poses as a lower limit for the mass of the wormlike chains.

5.4.7 Hydrodynamic Scale Relationships with Power Laws

A generalization of the hydrodynamic scale relationship underlying the determination of $D(s)$ is the power law

$$s = \kappa_s M^{b_s} \tag{5.13}$$

with constant coefficients [159]. It is identical to the $s \sim M^{2/3}$ power law for compact particles underlying the standard $c(s)$ relationship with $b_s = 0.66$ and $\kappa_s \sim 1/f_{r,w}$. Other exponents apply for different kinds of polymers, for example, 0.4–0.5 for coils and ~0.15–0.2 for rods [159]. Thus, the general scaling law model is very useful in the analysis of polysaccharides and other polymers [159].

For SV analysis in the context of $c(s)$, Eq. (5.13) can be combined with the Svedberg equation Eq. (1.35) to replace Eq. (5.9) with

$$D(s) = s^{1-1/b_s} \kappa_s^{1/b_s} \frac{RT}{1 - \bar{v}\rho}. \tag{5.14}$$

It seems impossible to extract both κ_s and b_s from experimental SV data to be analyzed. However, values for both may be derived from the s-value observed for "calibration" standards with given (average) molar mass, for example, from the slope and intercept of a plot of $\log s$ vs $\log M$. In particular, this seems essential for the exponent b_s, whereas the coefficient κ_s may be determined by refinement in nonlinear regression (analogous to $f_{r,w}$).

> In the SEDFIT implementation, which is invoked with the function Continuous c(s) with general scaling law, only the coefficient κ_s can be refined, whereas the power coefficient b_s is fixed. This reflects the information content of typical SV data.

5.4.8 $c(s)$ with Tabulated Hydrodynamic Relationships

Finally, $D(s)$ may be defined freely on the basis of a tabulated relationship between s and M. This is useful if no clear hydrodynamic rule can be identified, but model material of a narrow mass range is available for standard experiments. Once $M(s)$ is empirically established at discrete points over the required range from s_{min} to s_{max}, it may be interpolated to any grid point of the distribution, and via the Svedberg equation the diffusion coefficient may be found as $D(s) = RTs/(M(s)(1 - \bar{v}\rho))$.

The corresponding SEDFIT function is Continuous c(s) with M(s) Table. It requires an ASCII file with a two-column table of s and M pairs, to be named mofs.dat and to be placed in the SEDFIT home directory.

5.4.9 Relationship between Different $c(s)$ Variants

It is quite obvious that in the limiting case of a mono-disperse distribution, the precise hydrodynamic scale relationship is irrelevant, as long as the parameter controlling diffusion is refined in the analysis. This is true generally for distributions that are not too broad, such that differences in $D(s)$ of the different models remain small.

To examine this point more quantitatively, let us consider the general scaling law and highlight the difference to the constant frictional ratio model: Rather than $s = \kappa_s M^{b_s}$, as in Eq. (5.13), we can write the same relationship identically using a mass-dependent pre-exponential factor $\kappa_{2/3}(M)$ and constant 2/3-power of globular molecules which is the basis for the constant frictional ratio model

$$s = \kappa_s M^{b_s} = \kappa_{2/3}(M)M^{2/3}, \tag{5.15}$$

where

$$\kappa_{2/3}(M) = \kappa_s M^{b_s - 2/3} \tag{5.16}$$

could be though of as a mass-dependent frictional ratio (*modulo* normalization factors). When using the $c(s)$ model with a single frictional ratio for a very narrow distribution that approaches a mono-disperse single-species model with modal mass M_0, the refinement of $f_{r,w}$ amounts to the determination of

$$\kappa_{2/3}(M_0) = \kappa_s M_0^{b_s - 2/3} =: \kappa_{2/3,0}. \tag{5.17}$$

In other words, the frictional ratio will adjust to that of the average species of the narrow distribution, determining $\kappa_{2/3,0}$. For distributions of finite widths we can use a Taylor series to express the relationship between s and M,

$$s = \left(\kappa_{2/3,0} + \left. \frac{d\kappa_{2/3}}{dM} \right|_0 \Delta M + \dots \right) (M_0 + \Delta M + \dots)^{2/3} \tag{5.18}$$

with the derivative

$$\left. \frac{d\kappa_{2/3}}{dM} \right|_0 = \kappa_s \left(b_s - \frac{2}{3} \right) \left. \frac{M^{b_s - 2/3}}{M} \right|_0 = \kappa_{2/3,0} \left(b_s - \frac{2}{3} \right) \frac{1}{M_0}. \tag{5.19}$$

Together, dropping quadratic and higher terms in ΔM, we have

$$s \cong \kappa_{2/3,0} \left[1 + \left(b_s - \frac{2}{3} \right) \frac{\Delta M}{M_0} \right] (M_0 + \Delta M)^{2/3} . \tag{5.20}$$

We can see from this that the leading term distinguishing the standard $c(s)$ with constant frictional ratio $f_{r,w}$ from the general scale relationship model is the second term in the brackets, growing only slowly with $\Delta M/M$, and with a rate dependent on the difference between the power coefficient and the value of $2/3$ for compact particles. Therefore, in a region surrounding M_0 the constant $f_{r,w}$ model will still offer a good description of the relationship between s and M.

As a consequence, different hydrodynamic scale relationships need to be considered only for broad distributions, or where they offer parameterizations particularly suitable to implement certain available prior knowledge.

5.5 THE MOLAR MASS DISTRIBUTION $c(M)$

Many questions that are addressed by SV experiments can be answered through determining a sedimentation coefficient distribution, frequently followed by a secondary analysis. Sometimes, however, the molar masses of the species or the molar mass distribution is the primary problem to be studied. With any of the variations for the scale relationship $D(s)$, the $c(s)$ distribution above can be immediately transformed to a molar mass distribution via the Svedberg equation. The new basis of the distribution must describe the same particles, therefore $c(s)ds = c(M)dM$, from which follows that we can calculate $c(M)$ simply by multiplication with a normalization function based on the derivative of $M(s)$,

$$c(M) = c(s) \times \left(\frac{dM}{ds} \right)^{-1} . \tag{5.21}$$

For a given scale relationship $D(s)$, we can transform this to $dM/ds = (dM/dD) \times (dD/ds)$, and

$$c(M) = c(s) \times \frac{D^2(1 - \bar{v}\rho)}{sRT} \times \left(-\frac{dD}{ds} \right)^{-1} . \tag{5.22}$$

Per definition — analogous to the definition of the sedimentation coefficient distribution Eq. (3.2) — the integral of the molar mass distribution over the complete distribution gives the total loading concentration

$$\int_{M_{\min}(s_{\min})}^{M_{\max}(s_{\max})} c(M)dM = c_{0,\text{tot}} , \tag{5.23}$$

and likewise the integral over a specific range of M-values gives the loading concentration of sedimenting material in the respective molar mass interval.

Unfortunately, this transformation Eq. (5.22) is nonlinear and introduces a dependence of the peak positions of $c(M)$ on the diffusion coefficients. Whereas in

$c(s)$ it was possible to consider $D(s)$ an operational approximation for diffusional deconvolution that allowed us to increase the resolution of sedimentation coefficient distributions over the ls-$g^*(s)$ approach, the physical interpretation of diffusion coefficients is central to $c(M)$. Therefore, in contrast to the $c(s)$ model, where the degree of diffusional deconvolution determines just the peak width while leaving the peak positions largely invariant, in the $c(M)$ model both the peak resolution and position are affected by the assumptions on $D(s)$.

For example, when using the most common $c(s)$ variant based on the scaling law for compact particles, different values for the average frictional ratio $f_{r,w}$ will generally leave major peaks invariant in position by $c(s)$, but lead to different peak M-values in $c(M)$. Nevertheless, if there is only one major peak per $c(s)$ segment and the experimental conditions are such that the best-fit $f_{r,w}$-value is well determined, the molar mass estimates are typically within 5–10% of the true value. If there is not only one but two dominant peaks, then the peak $c(M)$-values may still be good estimates if both underlying species exhibit the similar frictional ratios, which may or may not be the case. Otherwise, if the species represented in these peaks exhibit different frictional ratios, the best-fit frictional ratio $f_{r,w}$ will not reflect that of any of the two species, but an average. As a consequence, the molar mass of one species will be overestimated, and that of the other species will be underestimated. Therefore, when molar mass values are of interest for samples showing multiple major $c(s)$ peaks, a refined analysis should be performed using the segmented $c(s)$ or the hybrid discrete/continuous models to ensure each species of interest is characterized by its own $f_{r,w}$-value or discrete species.

An overview and detailed comparison of the many different methods for determining species molar masses can be found in Section 8.3.4. Briefly, the molar mass estimates from $c(M)$ analysis of SV experiments are far superior to those from discrete Lamm equation fits, either in the raw data space (compare Fig. 3.3 on p. 59) or with the Gaussian approximations in the space of apparent sedimentation coefficients (compare Fig. 4.6 on p. 90). Because of the intrinsic ability to account for sample heterogeneity and micro-heterogeneity, both within the main sedimentation boundary as well as in trailing or leading boundaries of species sedimenting slower or faster, the overall boundary broadening will not be mistaken for diffusional spread of a single species.[15]

For example, for the system introduced in Fig. 3.2C, consisting of 60% of a 100 kDa species in superposition with an 80 kDa species and a 120 kDa species, both at 20% of the total signal, the single species fit resulted in a molar mass estimate of 66 kDa (Fig. 3.3). By contrast — despite the fact that the underlying species cannot be resolved — integration of the $c(M)$ distribution leads to an average of 101 kDa, with the shape of $c(M)$ clearly revealing the heterogeneity (Fig. 5.11).

---

[15]Further, from its origin in the hydrodynamic scaling law, $c(M)$ by design avoids hydrodynamically impossible (s/M) pairs, in contrast to the discrete species models.

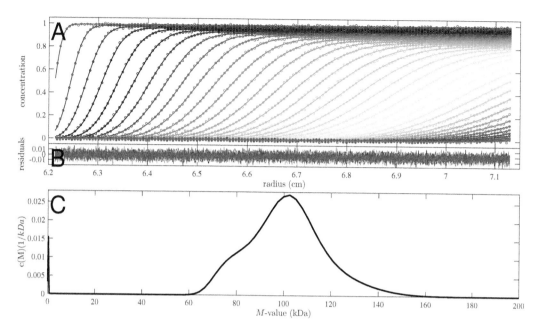

Figure 5.11 $c(M)$ analysis of the heterogeneous mixture from Fig. 3.2C, consisting of 60% of a 100 kDa, 6 S species in superposition with an 80 kDa, 5 S species and a 120 kDa, 7 S species, both at 20% of the total signal, with normally distributed noise at 0.005. *Panel A*: The fitted sedimentation data (circles, only every 5th data point shown) and best-fit profiles (solid lines), with residuals plotted in *Panel B* and the corresponding best-fit $c(M)$ distribution in *Panel C*.

SEDFIT offers different ways to determine molar mass distributions $c(M)$. (1) Prior to any data analysis, $c(M)$ models can be selected directly in the `Model` menu using the functions `Continuous c(M) Distribution` or those in the submenu `Continuous c(M) with other prior knowledge`. (2) Alternatively, any currently active $c(s)$ model can be converted into its corresponding $c(M)$ with the transformation found in the `Options ▷ Size Distribution Options ▷ transform c(s) to c(M)`. (3) An currently active $c(s)$ fit can be transformed to $c(M)$ by clicking on the corresponding model in the `Model` menu. A `Run` command should be issued to recalculate the distribution, forcing the renormalization of peak integrals *via* Eq. (5.23). Mass information can also be obtained for a $c(s)$ peak after integration.

5.6 LIMITATIONS OF DIFFUSIONAL DECONVOLUTION OF LARGE PARTICLES WITH ANY METHOD

Very large particles may not diffuse significantly over the time course of sedimentation due to their low translational diffusion coefficient. In the limiting case, the distribution will be governed completely by size heterogeneity. There is an intermediate size range where the distinction between diffusion and size heterogeneity is not possible on the basis of the experimental sedimentation data. If diffusion is not significant, then no deconvolution should be attempted.

A first, simple test is the comparison of the rmsd of the best-fit $c(s)$ distribution and with the best-fit ls-$g^*(s)$. If accounting for diffusion in the $c(s)$ distribution improves the fit beyond a statistically significant threshold as prescribed by F-statistics [71, 132], then diffusion can and should be deconvoluted and $c(s)$ should be used. If not, then deconvolution diffusion may introduce artificial features and ls-$g^*(s)$ should be used. If diffusion is significant, as a safeguard against overinterpretation of the data, it is advisable to probe the statistical confidence of the diffusion parameter (e.g., $f_{r,w}$) and corresponding error propagation into the features of the distribution. If diffusion is significant but $f_{r,w}$ is not well determined by the data, it may be assigned a value that is reasonable for the sample under study, as a first approximation of diffusion deconvolution.

Further, it can be very instructive to visualize the predicted diffusional boundary spread in the raw data space.[16] This may be accomplished, for example, by choosing a very coarse discretization of the distribution, i.e., a low N-value, and issuing a Run command to generate the boundary model. The goal is to make the distribution sufficiently coarse to display the different steps corresponding to different s-values. The width of these steps can then be visually compared with the width of the sedimentation boundary. If the theoretically predicted diffusion broadening is much less than the boundary spread, then polydispersity is the dominant factor, and any attempt to derive diffusion information from the data is questionable. Related, if diffusion coefficients are available, for example, from dynamic light scattering experiments, it is also possible to visualize the extent of diffusional spread in the raw data space after fixing D to this value in the SV analysis.

Known diffusion coefficients can be introduced in the simulation of boundary spread in the raw data space in the Non-Interacting Discrete Species model, after toggling the model to parameters s and D with Options ▷ Fitting Options ▷ Fit M and s. The comparison with experimental sedimentation boundaries is easiest when assigning the average s-value to the discrete species.

5.7 BAYESIAN ANALYSIS

The need for regularization was briefly introduced in Section 3.3 as a general characteristic of any distribution analysis: From the family of distributions that fit the data statistically indistinguishably well we have to identify the simplest one, in order to avoid noise amplification that would occur in a naïve best-fit approach. Following Occam's razor, we have accepted the "simplest" distribution to be the most parsimonious one, which led us to Tikhonov regularization and maximum entropy regularization. Here we will discuss how regularization can be refined using prior knowledge of the distribution in order to increase sensitivity and resolution. These concepts apply equally to standard $c(s)$ analysis and to size-and-shape distributions, and multi-component MSSV and MCMC analyses.

---

[16]If applicable, after subtraction of systematic noise components.

Let us consider the peak shapes that we would obtain for the same sedimentation process if we were to observe it with lower and lower signal/noise ratio. As illustrated in Fig. 5.12, the $c(s)$ analysis is able to provide clear resolution of two species sedimenting at 3 S and 5 S data from sedimentation data acquired at 50,000 rpm with a signal/noise ratio of 10:1. However, at a lower signal/noise ratio the peaks get broader. At a signal/noise ratio of 3:1, only the approximate range of s-values in the sample can be discerned, whereas at a signal/noise ratio of 1:1 very little can be said other than that there is something sedimenting.[17] Interestingly, even for data consisting entirely of noise, the $c(s)$ distribution is not zero when using maximum entropy regularization, but remains at a constant small value.

The reason for this broadening of the peaks is, of course, the regularization. Computationally, as described in Eq. (3.6) of Section 3.3, we do not want to calculate strictly the best-fit distribution by finding the lowest sum of squares (for which we will use the abbreviation $\Delta_{\text{fit}}^2(\gamma(s))$ in the following), since this approach would render the best-fit distribution extremely susceptible to experimental noise (Fig. 3.8). Rather, we perform a constrained optimization

$$\underset{\gamma(s)\geq 0}{Min}\ \left(\alpha\Omega[\gamma(s)] + \Delta_{\text{fit}}^2(\gamma(s))\right), \tag{5.24}$$

where we try, simultaneously with fitting the raw data, to also optimize $\Omega[\gamma(s)]$ as much as possible. $\Omega[\gamma(s)]$ may be a measure of smoothness of the distribution $\gamma(s)$ in Tikhonov regularization Eq. (3.8) or its information entropy in maximum entropy regularization Eq. (3.9). These measures of information content are scaled via proper choice of α to the maximum value not yet causing a significant decrease of the quality of fit. Consequently, as the signal/noise ratio decreases, the regularization term becomes more and more relevant. This is because at lower information content of the data, the threshold for the regularization to impose a statistically significantly worse fit is becoming higher.

At the very lowest signal/noise ratio (red line in Fig. 5.12), the resulting distribution is close to a flat line, which contains no information about different s-values. This makes sense, since if there is no signal in the data and only noise, we should not be misled to draw any conclusions from the experiment.[18] Conversely, the better the signal/noise ratio, and the more informative the experimental conditions, the less "room" there is for the regularization, the sharper the peaks, and the better we can determine the s-values of all species.

Unfortunately, by limiting the information in $c(s)$ to the minimum necessary for explaining the data, the standard regularization also highlights open questions:

---

[17]The presence of sedimenting material can be concluded from the comparison of the rmsd of a fit allowing for sedimentation with that excluding any sedimentation. However, the detailed signal/noise ratio where this behavior sets in depends on the noise model, number of scans, density of radial points, and other factors.

[18]With the naïve analysis in this situation we would not notice that lack of information: Rather than broad peaks we would still obtain trains of sharp peaks, but without any feature that would be statistically significant.

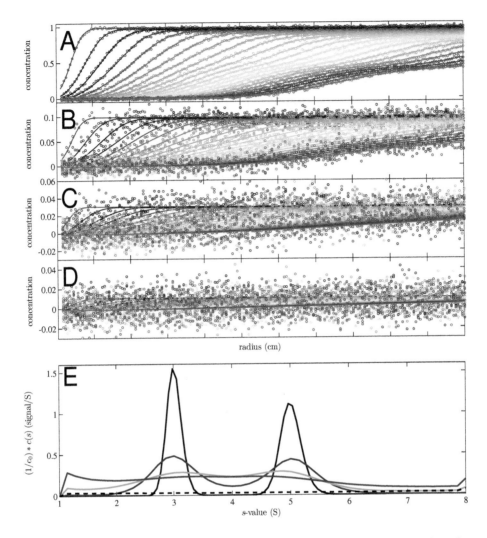

Figure 5.12 Illustration of the effect of the signal/noise ratio on the peak shape and resolution of the $c(s)$ distribution. SV data were simulated for two species at 3 S and 5 S at 50,000 rpm with signal/noise ratios of 100 (A), 10 (B), 3 (C), 1 (D), and 0 (not shown). The circles represent every 5th data point of every 2nd scan. The data are subjected to a $c(s)$ analysis without TI and RI noise and maximum entropy regularization is applied at a level of P = 0.68. *Panel E*: The resulting $c(s)$ distributions normalized for signal/noise ratios of 100 (dark blue), 10 (magenta), 3 (cyan), 1 (red), and 0 (black dashed line, using the same normalization factor as the red trace).

Since the peak shapes are now determined not only by the sedimenting species, but also by the signal/noise ratio and "information content" of the SV experiment, how can we decide whether or not two sedimentation coefficient distributions are from the same underlying distribution of particles? And how can we bring expectations from previous SV experiments into consideration in the interpretation of low information content data? These questions can be addressed by a refined, Bayesian approach of regularization [58, 169].

In this technique, rather than selecting the overall most parsimonious distribution with minimal information, we can establish a prior probability $p(s)$, and calculate the most parsimonious distribution *relative to this prior probability*. If we use the symbol $\Omega[\gamma(s), p(s)]$ for the regularization functional given the prior probability $p(s)$, instead of Eq. (5.24) we can write the new analysis as

$$\underset{\gamma(s) \geq 0}{Min} \left(\alpha \Omega[\gamma(s), p(s)] + \Delta_{\mathrm{fit}}^2(\gamma(s)) \right) . \tag{5.25}$$

Specific implementations for $\Omega[\gamma(s), p(s)]$ can be

$$\Omega[c] = \int \left(\frac{d}{ds} \left(\frac{c(s)}{p(s)} \right) \right)^2 ds \tag{5.26}$$

in extension of Eq. (3.8) for Tikhonov–Phillips regularization,[19] and

$$\Omega[c] = \int c(s) \log \frac{c(s)}{p(s)} ds \tag{5.27}$$

for maximum entropy regularization, extending Eq. (3.9) [58]. The effect of this modification is that we now obtain the sedimentation coefficient distribution that deviates *as little as possible, but as much as necessary, from the prior probability distribution in order to maintain a fit quality statistically indistinguishable from the overall best fit*. This new choice is illustrated in Fig. 3.7 alongside the standard regularization, as a new direction in which we can probe the space of distributions that fits the data indistinguishably well. In the case of the sedimentation coefficient distribution $c(s)$, we refer to the result as $c^{(p)}(s)$ with the superscript indicating its origin of the Bayesian analysis using a prior probability.

If we choose $p(s) = const.$ we are at the starting point of the standard regularization that assigns equal probability to all s-values. To exploit the power of the Bayesian approach, we have to specify a distribution $p(s)$ that is more structured and embeds our prior knowledge or prior expectation. There are many possible sources for $p(s)$, including constructs assembled from Gaussians or discretized δ-functions positioned to reflect known s-values, $c(s)$ distributions obtained previously or with a different optical system, or theoretical predictions such as Gilbert–Jenkins theory [58]. In either case, it is very important to note that — in contrast to embedding prior knowledge as fixed constraints in the model — we cannot diminish the quality of fit by adding the prior expectation: the resulting $c^{(p)}(s)$ distribution will report features that may neglect or even contradict the prior expectation, if

---

[19]By minimizing the change in the ratio between the distribution and the prior, we make the choice of prior independent of the total loading concentration. The assignment of α, the overall amplitude of the prior probability distribution, is a difficult problem that cannot be easily solved theoretically. Part of the problem is that it is difficult to quantitatively express our degree of belief that certain features should occur. Fortunately, it is straightforward to work out empirically which amplitudes lead to distributions that are appreciably different from those of the standard regularization. Since all $c^{(p)}$ distributions by design lead to the same quality of fit as the standard regularization, choosing different amplitudes for the prior simply amounts to exploring the confidence interval in the space of sedimentation coefficient distributions.

warranted by the information content of the data.[20] As a consequence, we may sometimes simply use the Bayesian approach to generally probe the confidence interval of the distribution in function space. Instead of having true prior knowledge we may just hypothesize some particular property of the distribution and examine the consequence of this in the sedimentation coefficient distribution.

As a first illustration of this powerful concept, let us go back to the $c(s)$ distributions calculated with maximum entropy regularization from the data shown in Fig. 5.12. For example, if we knew from other, prior information that there should very likely be a species with 3 S, this can be utilized to make a distribution for the prior expectation $p(s)$ that consists of a uniform baseline with a sharp peak at 3 S. The resulting distributions, using the same data as before, are shown in Fig. 5.13A. Even though the prior knowledge only specified the 3 S peak, we learn from the data that there is another species (which is indispensable for getting a statistically acceptable fit). Even from the most noisy data, we can conclude from $c^{(p)}(s)$ that the known 3 S species alone cannot explain the data, but that there must additionally be a species present in solution that sediments distinctly faster.

The question if two SV data sets can be described with the same distribution can be answered with the Bayesian $c^{(p)}(s)$ analysis by taking a previously calculated $c(s)$ distribution to initialize the prior expectation $p(s)$. This question may arise, for example, in the analysis of an experiment conducted with interference and absorbance data acquisition. Due to the higher signal/noise ratio of the interference data, usually the resolution in $c(s)$ achieved with the interference data is better, and one may wonder if the absorbance signal is simply redundant to the interference signal, or if it detects some additional species (e.g., nucleotides or oligonucleotides having high extinction), or perhaps fails to detect other species (e.g., polymers without extinction). Another example for such an application of Bayesian analysis is the study of a concentration series of samples, such as commonly employed in the study of potentially interacting systems, where it is important to decide whether the data at low signal/noise ratio reveal dissociation of complexes. Here, too, one could take the calculated $c(s)$ distribution from the highest concentration (and highest signal/noise ratio) to initialize the prior expectation $p(s)$ in the analysis of the lower concentrations. When this strategy is applied to our standard data of Fig. 5.12, the result is as shown in Fig. 5.13B: After scaling proportional to the loading concentration, all distributions virtually superimpose with the $c(s)$ distribution from the highest concentration. This is not a surprise, as the data was simulated with the same two species, but it demonstrates how the correct prior can avoid noise-driven peak broadening.

Interestingly, on closer inspection, the amplitudes of the $c(s)$ peaks from the lowest concentrations (red and cyan) are noticeably higher than one would assume on the basis of the loading concentrations, and even the data consisting of pure noise (dashed black line) can assume this peak structure. The increased relative amplitude

---

[20]In fact, $c^{(p)}(s)$ will perfectly adhere to $p(s)$ only in the rare case of a purely confirmatory experiment, or a data set that consists entirely of noise.

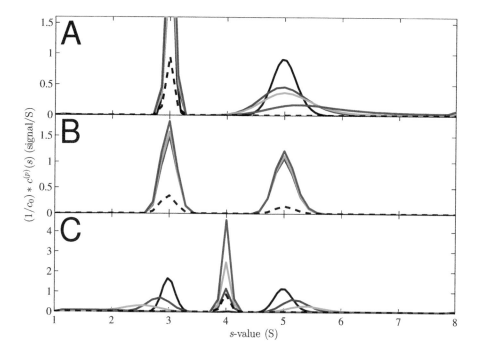

Figure 5.13 Analysis of the same data as in Fig. 5.12 modifying the $c(s)$ analysis with Bayesian priors under otherwise identical conditions. Traces are shown normalized as in Fig. 5.12 and in the same color scheme for signal/noise ratios of 100 (dark blue), 10 (magenta), 3 (cyan), 1 (red), and 0 (black dashed line, using the same normalization factor as the red trace). *Panel A*: The prior knowledge is used that we know a 3 S species exists, using $p(s) = \delta(s - 3)$. *Panel B*: The $c(s)$ distribution from the highest concentration is used as prior. *Panel C*: An impostor prior is used, incorrectly hypothesizing the presence of a 4 S species.

of the low signal/noise data is a result of the higher relative contributions from the noise. That the pure noise signal can be interpreted to some (small) extent as being caused by a distribution with such a peak structure is inherent in the random nature of the noise providing finite confidence limits for any model. Conceptually, this makes sense considering that without any new informative data, we cannot expect to learn anything new and contradictory to our expectations, and even within the degrees of freedom offered by the noise, we can recover our prior expectations. This excess amplitude drawn from the noise of the data produces a slight overestimate of the peak heights, which appears at higher relative contributions for the data sets with the low signal/noise ratios. A similar situation occurs in the standard regularization, where the signal from the noise or a trace species can be drawn to facilitate broader major $c(s)$ peaks [59].[21]

---

[21]For this reason, a modification of the standard regularization is advantageous for the application to trace determination of oligomeric aggregates [59, 170], where otherwise a fraction of the signal from a dimer will be mis-attributed to the monomer in order to make the monomer peak broader. See Section 8.2.5.

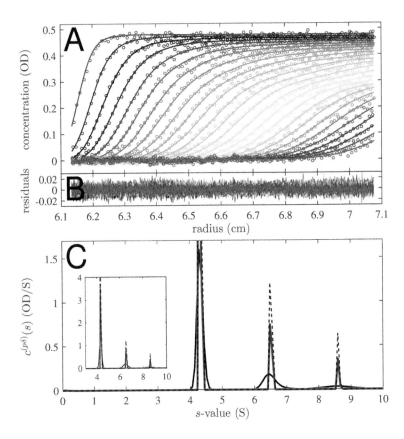

Figure 5.14 Analysis of SV data of BSA by $c^{(p\delta)}(s)$. *Panel A*: Absorbance SV data are acquired at 50,000 rpm and 20°C. The circles represent every 3rd data point of every 3rd scan. An initial $c(s)$ fit was carried out and peaks were converted to δ-functions maintaining s_w and peak area constituting the prior $p(s)$ for the following $c^{(p\delta)}(s)$ fit (colored solid lines in *A*) *Panel B*: Residuals of the final fit. *Panel C*: Initial standard $c(s)$ distribution (black solid line), the derived set of δ-functions used as $p(s)$ (blue dashed line) and the $c^{(p\delta)}(s)$ distribution (red). The inset shows the same graph in full scale with un-truncated $c^{(p\delta)}(s)$. Only 2.6% of total concentration reflected in $c^{(p\delta)}(s)$ is not consistent with the δ-function prior.

An important question is: What happens when our prior expectation is wrong? For our standard example from Fig. 5.12, this can be examined by imposing the impostor prior embedding the expectation that the distribution should resemble a single, sharp peak at the s-value of 4 S. The resulting distributions are shown in Fig. 5.13C. Since the information content of the data clashes with the prior expectation, for the data at high signal/noise ratio (dark blue line), only a very small 4 S peak can be accommodated by the data (allowed for by the finite noise in the data), while the majority of the material is correctly identified as sedimenting with 3 S and 5 S, respectively. This highlights the gentleness of the Bayesian approach of prior expectation, as opposed to a hard built-in constraint of the model that would not allow for anything but a 4 S species resulting in a poor fit. For the data at a smaller signal/noise ratio of 10:1, a portion of the signal can be attributed to the false 4 S peak, but in order to achieve a good fit side lobes in the range between 1.0 S and 2.5 S and between 5.5 S and 7 S appear, which reflect the true species. Only in the absence of any sedimentation signal, pure noise can be interpreted to be entirely consistent with a very small population of the false species.

When working with biological samples, a very useful type of prior knowledge can be derived from the fact that they can frequently be prepared as mono-disperse ensembles of macromolecules with uniform size and shape,[22] although usually additionally with some degradation products and impurities. Since the degradation and aggregation products can usually be expected to sediment at different s-values, often a reasonable prior expectation would be that the main peaks reflect separate single species. In this case, the prior $p(s)$ can be bootstrapped from an initial standard $c(s)$ analysis, followed by the automatic replacement of all peaks by the best possible discretized approximation of δ-functions that maintain the s_w and area of each peak. A new distribution $c^{(p)}$ is then calculated using this prior. To indicate this particular strategy to use δ-function priors to express the expectation of single species, the resulting distribution is termed $c^{(p\delta)}$ [58].

Bayesian regularization can be switched on in SEDFIT with the function Options ▷ Size Distribution Options ▷ Use Prior Probabilities. This leads to an input box allowing the user to select alternative sources for $p(s)$:

(1) An arbitrarily defined prior from a file to be specified after closing of the box. The file is assumed to be a standard two-column ASCII format, a format also used, e.g., in the file generated by Data ▷ Save Continuous Distribution. The s-range in $p(s)$ file is expected to be overlapping the currently selected $c(s)$ range;

(2) Select Peaks From Current Distribution uses the automatically bootstrapped δ-functions preserving s_w and peak areas from an existing current $c(s)$ distribution as $p(s)$;

(3) An assembly of manually entered Gaussians of certain center and width (along with an arbitrary amplitude factor that can be empirically adjusted so as to make the Bayesian prior effective).

The same options exist for the size-and-shape analysis, appended by the input of the prior in the f_r dimension. Selection of the $c^{(p\delta)}$ model in this box is equivalent to the separate Size Distribution Options menu function Prior Knowledge of Discrete Species, which can be automatically executed from an existing $c(s)$ analysis with the keyboard shortcut control-X.

The use of Bayesian regularization for MSSV analysis can be signalled by pressing the push-buttons PP adjacent to each branch of each s-segment. After closing the box, a series of Gaussians can be entered to serve as priors for each branch.

The selection of any of these prior models must be followed by a RUN command to carry out the new analysis. It is not necessary to do a FIT, since the prior will have no effect on the non-linear parameters of the model.

---

[22] As mentioned previously, conformational fluctuations will not be visible in the sedimentation experiments if they are fast relative to the time-scale of the experiment, i.e., have a life-time less than $\sim$100 sec (Part I, Chapter 2).

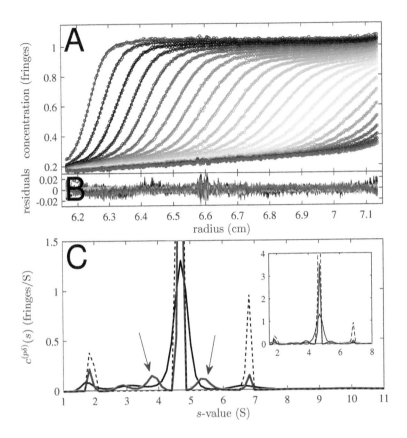

Figure 5.15 Analysis of SV data of a heavily glycosylated NK receptor fragment by $c^{(p\delta)}(s)$. *Panel A*: Interference SV data are acquired at 50,000 rpm and 20°C. The circles represent every 5th data point of every 3rd scan. As in Fig. 5.14, an initial $c(s)$ fit was carried out and peaks converted to δ-functions constituting the prior $p(s)$ for the following $c^{(p\delta)}(s)$ fit (colored solid lines) *Panel B*: Residuals of the final fit. *Panel C*: Initial standard $c(s)$ distribution (black), the derived $p(s)$ (blue dashed) and $c^{(p\delta)}(s)$ distribution (red). The inset shows the same graph in full scale with un-truncated $c^{(p\delta)}(s)$. In contrast to Fig. 5.14, 28.4% of total concentration reflected in $c^{(p\delta)}(s)$ is not consistent with the δ-function prior. Most importantly, due to the "compression" of the main peak into a δ-function, new side lobes (arrows) arise that ensure the quality of fit does not suffer.

Fig. 5.14 illustrates this approach for a sedimentation experiment with BSA. The initial $c(s)$ distribution is shown as a solid black line. Automated integration of each peak produced the set of δ-functions used as the prior, shown as a blue dashed line. The resulting $c^{(p\delta)}(s)$, shown in red, follows the expectation that each major peak represents — at least for the purpose of the SV experiment — a distinct, mono-disperse state. By contrast, Fig. 5.15 shows an example of $c^{(p\delta)}(s)$ applied to a sample where the main species is not mono-disperse, namely a heavily glycosylated NK receptor fragment [67]. In this case, the broad main peak cannot follow a δ-peak without causing side lobes to emerge in $c^{(p\delta)}(s)$. These show that the assumption that the main peak in $c(s)$ can be substituted by a single species is wrong, or at least incomplete. These new side peaks in $c^{(p\delta)}(s)$ could, in principle, represent existing contamination with slightly slower and slightly faster sedimenting species. On the other hand, they could also be compensatory peaks for the truly more polydisperse ensemble of molecules in this main peak, which would not be correctly captured in the prior of $c^{(p\delta)}(s)$. Which alternative is true cannot be decided from the SV data alone. However, considering that this molecule is heavily glycosylated — a process that frequently results in significant heterogeneity of the molar mass — one would tend to interpret the side lobes to be compensatory elements in the fit, and in this case dismiss the $c^{(p\delta)}(s)$ model in favor of the original $c(s)$ distribution. In this application, the exercise of $c^{(p\delta)}(s)$ has informed us that the main peak is not a single species, and that the polydispersity in glycosylation leads to a range of sedimentation coefficients.[23]

In order to facilitate the assessment to which extent the $c^{(p\delta)}$ distribution is consistent with the prior $p(s)$, SEDFIT will report a list of the δ-function peaks of $p(s)$ with their respective s_w-values, populations in signal units, and % total loading signal. The material sedimenting outside these pre-identified δ-functions is listed as OTHER MATERIAL in % units and roughly identified by the overall s_w of these contributions. Inspection of this table is highly useful since the sharpness of the peaks in $c^{(p\delta)}$ can make it difficult to visually discern broader and shallower peaks even if the latter describe significant fractions of the total material.

In summary, introduction of Bayesian principles into the regularization allows us to modulate the behavior of sedimentation coefficient distributions under low signal/noise conditions in a way that exactly maintains the quality of fit. Typically this can greatly improve the resolution and information content of the resulting distribution. Beyond the examples described above, other applications are possible, some of which are described in Section 8.2.5.

5.8 SUMMARY: LEVELS OF INFORMATION

As a summary of this chapter of methods for sedimentation coefficient distribution analysis for non-interacting mixtures, it is instructive to recapitulate the different methods in light of how much and what type of information they can extract from the experimental SV boundaries. Distribution analysis is essential for most SV experiments because of their exquisite sensitivity to heterogeneity. Unless the data fit satisfactorily to a single- or multiple-discrete-species model *over the entire time course of sedimentation*, which is quite rare, the distribution analysis is

---

[23]Theoretically, a third reason for the side lobes could be a significant underestimate of the diffusional spread in $c(s)$, which could arise from a significant overestimate of the frictional ratio when using the standard scaling law. In this case, a finite width in the $c(s)$ peak is necessary for mimicking unaccounted diffusion. This effect is minimized by ensuring a good quality of fit, and can be ruled out if the molar mass estimate associated with the peak is close (i.e., within 10%) to the known true value. Also, this situation is unlikely to occur if the peak in question is the only major peak in the particular $c(s)$ segment. These conditions apply to the example in Fig. 5.15.

indispensable for determining well-defined estimates of loading concentrations, sedimentation coefficients, and molar masses.

Assuming the SV data have a canonical form of migrating sedimentation boundaries and a model that fits the data well, and additionally assuming that we have verified through a concentration series that no attractive or repulsive interactions take place under the experimental conditions, then there is a natural hierarchy in the conclusions that can be drawn from various features of the data:

(1) Solely the fact that we achieve a statistically acceptable fit to the SV data, with any model, implies estimates for the constant and systematic baseline offsets. As a consequence, we can determine the total loading concentration of material. Further, a satisfactory fit of the raw data implies reliable values for the integrals under the sedimentation boundaries at all points in time, from which a well-defined weighted average s_w-value directly follows. Since it is based on the transport method, this s_w-value is completely independent of the physical meaning of the theoretical model that was used to achieve the fit to the experimental data.

(2) If the SV data show multiple clearly separating boundaries, their respective loading concentrations and sedimentation coefficients are also straightforward to determine with a distribution model.

(3) If we were to use a discrete species model, the boundary spread of a single boundary could be interpreted in terms of an apparent diffusion coefficient. *Via* the Svedberg relationship and using the s-value, it could be translated into an apparent buoyant molar mass value. In most cases, such an apparent diffusion coefficient is an overestimate, and the apparent molar mass value is an underestimate of the true molar mass, sometimes by a large margin, due to unresolved heterogeneity of sedimenting species often contributing to form this sedimentation boundary.[24]

(4) Resolving possible heterogeneity of the species represented in the boundary requires diffusional deconvolution. This can be achieved by considering experimental data over a large time span, which allows the distinction of components of the temporal evolution that proceed exponentially with time $\sim e^{s\omega^2 t}$ *vs.* those that exhibit a $\sqrt{t}$ time dependence. It is important to verify that a good fit is achieved. The most general model providing diffusional deconvolution is the size-and-shape distribution $c(s, f_r)$. It resolves populations of species with different s-values contributing to the same boundary, to the extent that their distinct signal contributions are essential to fit the data. Otherwise, it will provide a broad peak with the least amount of information necessary to interpret the data.

---

[24]If attractive or repulsive interactions are present under the experimental conditions, a useful apparent molar mass cannot be derived from the analysis of boundary spread.

(5) Diffusional deconvolution also reveals the presence of trace components in other regions outside the main boundaries. Usually, the amplitudes of these trace components are very well defined, but the s-value has a lower precision (compare Fig. 8.10 on p. 201).

(6) The discrimination of species with the same s-value but different diffusion coefficients (or frictional ratios, or molar masses, respectively) is usually difficult within the achievable signal/noise ratios of experimental SV data, and well-developed peaks in this dimension (allowing for good estimates of the molar mass) occur for the major species only. Therefore we may neglect this second dimension and reduce the distribution to the general $c(s, *)$. Alternatively, however, if we utilize additional knowledge about the nature of the sample and embed that into a scaling law $D(s)$, the resulting $c(s)$ distribution typically provides a better resolution, a more robust determination of trace species, and more reliable molar mass estimates. This is the strategy used in most applications.

(7) Instead of more knowledge, we can provide more data. This can be in the form of SV data acquired under conditions providing different ratios of sedimentation $vs.$ diffusion fluxes, or from different techniques such as SE or DLS. The global analysis will generally have a higher resolution of species and, in particular, more precise information on the molar mass of all species. However, such a global analysis requires the detailed understanding of all signal contributions and imperfections in each experiment, to ensure the data are compatible.

(8) Another variant of the strategy to offer more data specifically for multi-component mixtures is the inclusion of different optical signals. This leads to the multi-signal $c_k(s)$ that directly reports on the stoichiometry of complexes, from which the molar mass may also be deduced. Similarly, spectral discrimination may be substituted by characteristic temporal signal changes.

(9) Finally, we can achieve higher resolution in any of the sedimentation coefficient distributions if we contribute additional prior knowledge, which can be accomplished with the Bayesian approach of regularization in $c^{(p)}(s)$. This allows us to examine in more hydrodynamic detail the species that contribute to a single $c(s)$ peak of finite width.

As in any other biophysical discipline, understanding the strength and weaknesses of the different models, and mastering the flow of information — the give-and-take of knowledge and assumptions placed into the model and the response in terms of quality of fit and details of the results — is the essence of successful data analysis.

CHAPTER 6

Sedimentation Coefficient Distributions from Boundary Derivatives and Extrapolations

H ISTORICALLY, there has been intense and longstanding interest in the determination of particle size distributions, going back to the first description of the AUC by Svedberg and early applications [113, 171, 172]. Formal definitions of different distribution functions, including both sedimentation coefficient and molar mass distributions, can be found in communications from the 1950s, for example, by Williams, van Holde, Baldwin, and Fujita [111, 173, 174] and Gosting [175], who had embarked on the expansion of analytical approximations of the Lamm equation solutions to distributions, in largely theoretical work. Practical approaches were devised by several authors [110, 146], for example, via extrapolation schemes to address the problem of diffusion [110–112, 117, 146, 173, 175, 176]. Some of these methods were very sophisticated, but without the ability of solving the Lamm equation and fitting integral equations, they were based on concepts quite different from the modern ones.

This history is far beyond the scope of this work. However, we briefly review two methods that are the last examples of the idea to use data subset transformations and extrapolations for extracting information about the sedimentation coefficient distribution. Whether or not they are actually implemented using a computer as an aid, and whether or not they are embedded in modern software with some user convenience and automated features, they remain pre- or semi-computer approaches by nature, since they are based on simplified relationships in the spirit of addressing problems by plotting data in a certain way or using spreadsheet operations.

Discussion of these approaches in light of the framework developed so far provides additional examples for the application of theoretical relationships between the parameters governing the sedimentation process. Intimate familiarity with these methods is also very useful to interpret a part of the published literature where these methods have been or are still being used.

6.1 THE METHOD BY VAN HOLDE AND WEISCHET TO ESTIMATE THE INTEGRAL SEDIMENTATION COEFFICIENT DISTRIBUTION $G(s)$

A integral sedimentation coefficient distribution $G(s)$ reports on the s-value of a certain percentile of the total sedimenting material (ordered from lowest to highest sedimentation coefficient). It can be calculated best in relation to the diffusion-deconvoluted differential sedimentation coefficient distributions, alternatively as

$$G(s) := \int_0^s \int_1^\infty c(s', f_r) df_r ds' = \int_0^s c(s', *) ds' \qquad (6.1a)$$

$$G(s) := \int_0^s c(s') ds', \qquad (6.1b)$$

i.e., as an accumulative representation of either the size-and-shape distribution $c(s, f_r)$, or the sedimentation coefficient distribution $c(s)$ discussed above. In theory $G(s)$ is strictly equivalent in information content to its differential sedimentation coefficient distribution counterparts, although sometimes defined with a normalization constant $1/c_{tot}$.[1]

> The function Options ▷ Size Distribution Options ▷ transform differential to integral distribution in SEDFIT can calculate $G(s)$ by integration of a currently loaded differential sedimentation coefficient distribution model. If desired, a $G(s)$ representation can also be achieved in the graphing stage in the program GUSSI [177].

The van Holde–Weischet method [117] is an extrapolation scheme to approximate an integral sedimentation coefficient distribution $G(s)$ directly from sedimentation boundaries. Unfortunately, a strict interpretation as a real distribution of sedimenting species must often be abandoned, and an integral distribution of apparent sedimentation coefficients $G(s^*)$ would be a more appropriate, though

---

[1]In principle, therefore, a differential sedimentation coefficient distribution like $c(s, *)$ could be obtained from $G(s)$ by differentiation as $c(s) = (d/ds)G(s)$. However, in practice this is very problematic due to the noise in $G(s)$ being severely amplified in the differentiation process. Typically, for example, the $G(s)$ estimates resulting from the van Holde–Weischet approach are not monotonically increasing functions, such that differentiation could result in unphysical, negative concentrations in $c(s)$ or in multi-valued functions.

not conventional, nomenclature. Although the van Holde–Weischet method has become obsolete, it is based on approximations that are insightful and were applicable without the current abundance of computational resources. It is based essentially on the combination of two main ideas.

6.1.1 Boundary Divisions: Diffusion-Free Polydisperse Mixtures

The first seems pragmatic in nature and is a scheme to represent the boundary information in discrete steps. Historically, such a strategy was adopted in different forms by many groups as an efficient way to represent and analyze experimental boundary shapes at a time predating digital data acquisition and numerical computer methods [111, 117, 176, 178].

In the implementation of the van Holde–Weischet method, each sedimentation boundary at time t is measured from baseline to plateau and is then divided in N steps of equal height, indexed as $i = 1 \ldots N$, and their radial position $R_{i,t}$ is determined (Fig. 6.1). This divides up the boundary shape information into manageable pieces. The resulting radial positions are transformed into the space of apparent sedimentation coefficients via (2.3), following $s_{i,t}^* = (\omega^2 t)^{-1} \log (R_{i,t}/m)$.[2,3] Conceptually, the idea is that if the boundaries were from the diffusion-free ideal sedimentation of large particles, then the $s_{i,t}^*$-values for equivalent boundary fraction from each scan should be identical, all reporting on the s-values s_i of the species sedimenting in that boundary fraction i. As a consequence, the sequence of s_i-values for increasing boundary fractions would represent an approximation for an integral sedimentation coefficient distribution $G(s)$.

Although the boundary division is not commonly revealed in the raw data space, it is very instructive to visualize the boundary divisions as the sequence of discrete step functions to which the boundary division is equivalent. Intriguingly, the resulting "staircase" model of the boundaries is graphically reminiscent of the modern boundary modeling with the ls-$g^*(s)$ distribution (for example, compare Fig. 3.5 on

---

[2]In some instances, transformation from radius to the s^* axis preceded the boundary division, which is equivalent.

[3]Determining the radial positions $R_{i,t}$ is non-trivial due to the noise in the data. Avoiding bias from smoothing operations on the raw data, it is possible to determine $R_{i,t}$ as the mean of all data points with values within the limits of the boundary fraction [57]. The problem is exacerbated for shallow regions of the boundaries, such as the leading and trailing edges close to the plateaus: here, the error in the assignment of the radius is amplified with $\Delta R = \delta (da/dr)^{-1}$ (with da/dr being the gradient of the boundary and δ denoting the noise in the data). Thus, one would not expect high precision for trace components sedimenting slower or faster than the main peak.

Also, the method of boundary divisions is intrinsically incompatible with systematic radial-dependent TI noise, and therefore cannot be used for interference optical data. This may be remedied, in principle, by the acquisition of water blanks and subtraction of their profiles from the data, as discussed in Part I, Chapter 4, although the reliability of such a procedure has not been shown for SV data. *Ad hoc* operations to estimate the TI noise contribution from given SV data may often give acceptable results on a pragmatic level for the purpose of the van Holde–Weischet analysis.

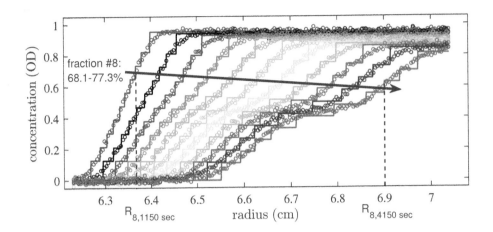

Figure 6.1 Boundary division in van Holde–Weischet analysis, illustrated with simulated data of a 100 kDa, 6 S species in a mixture with a 200 kDa, 10 S species sedimenting at 50,000 rpm. Shown are simulated signal profiles (circles) at 11 time points from 1150 sec to 4150 sec and boundary divisions of equal heights (solid lines), excluding half a step size at either plateau. Each step reflects the radial position $R_{i,t}$ of each scan covering a certain percentile of the total boundary height, as measured between solvent and solution plateaus. The radial positions shown are those obtained from the linear regression Eq. (6.4) as displayed in Fig. 6.2. As an example, the magenta arrow highlights the 8th fraction covering 68.1% to 77.3% of boundary heights.

p. 65). In fact, the detailed comparison provides an interesting application of the theoretical framework we have presented above, touching on several issues.

A fundamental difference is that in the graphical division the step heights are fixed and equal *a priori* and the radial positions are determined from the data, whereas in ls-$g^*(s)$ the radial positions are fixed (through the fixed s-grid and times of scans) and the step heights are fitted to the data. This is related to the fact that the former will provide an integral sedimentation coefficient distribution (asking "what is the s-value of a certain signal fraction?"), while the latter provides a differential sedimentation coefficient distribution ("what is the population of species with a certain s-value range?").

Since the division of the boundary takes place relative to the plateau height, the information about radial dilution and mass conservation is abandoned up front. As a consequence, the van Holde–Weischet method is fundamentally incompatible with the transport method and not suitable for the determination of a weighted-average s_w-value [107]. Another consequence of the boundary divisions being scaled to the plateau is that it can only be applied to a data subset that actually shows plateau regions. Usually, this significantly constrains the time range of the experimental data basis. In particular, this severely limits the application to broad distributions (where the fastest species shrink the radial range of the solvent plateau rapidly) and distributions including small molecules (which will diminish the plateau region quickly due to back-diffusion).

Scaling of the boundary division relative to the plateau height is aimed at ensuring that the same boundary fraction will represent the same subpopulation of sedimenting material. Although this may sound plausible at first sight, this is not really the case [57, 117]: As a consequence of the radial dilution being dependent on the species' s-values, the signal amplitudes of faster sedimenting species will decrease more strongly than those of the slowly sedimenting material. Scaling relative to the overall total boundary height will therefore result in the faster sedimenting species continually being compressed more and more into the upper boundary fractions. This mis-assignment is intrinsic to the application of the van Holde–Weischet method to heterogeneous mixtures and causes apparent populations with artificially intermediate s-values, as well as slight distortions in the represented relative amounts of different species. By contrast, the steps in the ls-$g^*(s)$ naturally obey the square dilution rule, not causing this error.[4]

Finally, whereas the steps in ls-$g^*(s)$ are determined in a global fit to all experimental boundaries that enforces them automatically to be consistent from one scan to the next, the boundary division scheme adopted in the van Holde–Weischet method treats each scan separately and subjects the $s_{i,t}^*$-values to a secondary analysis. Though not optimal, this was the best possible approach at the time, given the available computational resources.

6.1.2 Extrapolation to Infinite Time for Eliminating Diffusion of a Single Species

The analysis of $s_{i,t}^*$-values is the second stage of the van Holde–Weischet method. Its goal is to address the problem of diffusional transport superimposed on the sedimentation. Fractions in the leading edge of the diffusion-broadened boundary will be at higher radii than the boundary mid-point and therefore produce higher $s_{i,t}^*$-values, and fractions from the trailing part will produce lower $s_{i,t}^*$-values. Since the diffusional spread increases with time on a $\sqrt{t}$ scale, whereas displacement from sedimentation (or differential sedimentation) grows exponentially, in the limit of infinite time the $s_{i,t}^*$-values will converge to s_i^* [112]. The idea underlying the van Holde–Weischet method is the extrapolation of $s_{i,t}^*$ to infinite time by linear extrapolation of the plot of $s_{i,t}^*$ vs $1/\sqrt{t}$ to $1/\sqrt{t} \to 0$.[5]

---

[4]A more elaborate scheme of boundary divisions has been reported by Demeler and van Holde in 2004 to address this problem [179]. In brief, an iterative correction for radial dilution is nested into the extrapolation step determining s_i, preceded by an empirical polynomial extrapolation of the plateaus. The method is further enhanced by an innovative approach for differentiation and smoothing of $G(s)$ (consisting of the transformation of $G(s)$ into a histogram subsequently represented by Gaussian sums with *ad hoc* user-selected width), to arrive at an appearance of the familiar differential sedimentation coefficient distribution [179], with user-selected level of detail. The advantage of this enhanced van Holde–Weischet method is considered its rigor and absence of user bias [179].

[5]With regard to the history of this extrapolation, contrary to the suggestion by van Holde and Weischet in their 1978 paper [117], the introduction of boundary fractions by Gralén and Lagermalm [178] in 1952 was not for the purpose of carrying out an extrapolation of boundary fractions to infinite time $1/t \to 0$. In fact, under the experimental conditions, diffusion was considered to be negligible for the large polystyrene molecules studied, as described in their paper [178]. An

In quantitative terms, the theoretical basis of the approach is the Faxén-type approximate solution of the Lamm equation introduced in Eq. (2.31). As outlined by van Holde and Weischet [117], when boundary divisions are applied to these approximate solutions for a single species sedimenting with s_i, the fractional concentrations $\chi(R_{i,t}, t)$ scaled relative to the plateau concentrations are constant i/N, such that

$$\frac{i}{N} = \frac{1}{2}\left[1 - \Phi\left(\frac{m\left(\omega^2 s_i t - \log\left(R_{i,t}/m\right)\right)}{2\sqrt{Dt}}\right)\right] \tag{6.2}$$

where $\Phi(x)$ is the error function $(2/\sqrt{\pi})\int_0^x e^{-t^2}\,dt$. After transformation to the s^*-scale via Eq. (2.3), we have

$$\frac{2i}{N} = 1 - \Phi\left((s_i - s_{i,t}^*)\frac{m\omega^2}{2\sqrt{D}}\sqrt{t}\right). \tag{6.3}$$

The key analytic operation is the use of the inverse error function $\Phi^{-1}(y)$, for which $\Phi^{-1}\Phi(x) = x$ holds. Rearranging, we arrive at

$$s_{i,t}^* = s_i - \frac{2\sqrt{D}}{m\omega^2}\Phi^{-1}\left(1 - \frac{2i}{N}\right)\frac{1}{\sqrt{t}}. \tag{6.4}$$

This describes a linear function of $s_{i,t}^*$ vs $1/\sqrt{t}$ that converges $s_{i,t}^* \to s_i$ as $1/\sqrt{t} \to 0$ [117].[6] This satisfactory result motivates the plot of $s_{i,t}^*$ on a $1/\sqrt{t}$ scale and the linear regression of equivalent boundary fractions to determine s_i, which can be carried out using analytical expressions. The sequence s_i represents the inverse sedimentation coefficient distribution, $G(s_i) = c_0\, i/N$.[7] For the data of Fig. 6.1 this operation is shown in Fig. 6.2.

---

extrapolation was carried out across equivalent boundary fractions from samples at different concentrations, to extrapolate the sedimentation coefficient distribution to infinite dilution. Diffusion was then studied in separate experiments using the total area/maximum height ratio of the diffusion curve.

A temporal extrapolation was introduced in 1950 by Baldwin and Williams [110] and in more detail in 1959 by Baldwin [111] for Gaussian sedimentation coefficient distributions $g(s)$. Baldwin considered the values of differential sedimentation coefficient distributions at fixed values of $g^*(s)/g_{max}^*(s)$, which corresponds approximately to equivalent concentration fractions. With the goal of eliminating the effect of diffusion on the sedimentation coefficient distribution of polydisperse samples, an extrapolation, $(\bar{s} - s^*)^2$ vs. $e^{-\bar{s}\omega^2 t}/t \to 0$ was applied. Later, in 1977, Winzor and colleagues [176] proposed the extrapolation of $s_{i,t}^*$-values of boundary fractions to infinite time on the basis of $1/t \to 0$, for the purpose of determining asymptotic shapes of reaction boundaries.

[6]Neither Φ nor Φ^{-1} actually needs to be evaluated to calculate $G(s)$, since the error function determines solely the slope of the straight line. If the latter should be interpreted in terms of a diffusion coefficient, a table of Φ^{-1} needs to be consulted. However, van Holde and Weischet noted that diffusion coefficients determined with this method will not be very accurate [117].

[7]As a quality control, it is possible to back-project the best-fit $s_{i,t}^*$-values from the linear regression into the raw data space as steps at $R_{i,t}$. However, in the original pre-computer approach, this was not done.

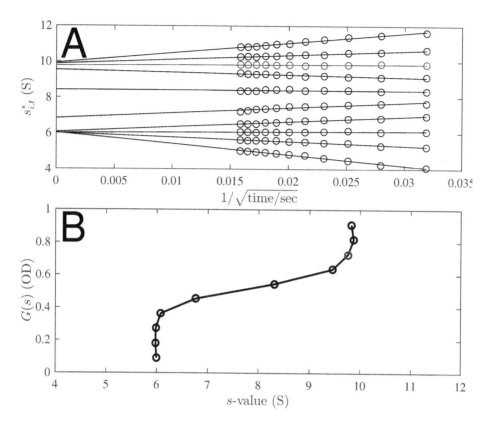

Figure 6.2 Principle of the extrapolation in van Holde–Weischet analysis, illustrated with data from the two-species data of Fig. 6.1. *Panel A*: Plot of the boundary positions $R_{i,t}$ after transformation to apparent s-values (circles) on the $1/\sqrt{t}$ scale . The solid lines are best linear fits to each boundary fraction. *Panel B*: The signal fraction of each boundary fraction is plotted *versus* the s^*-value of the same boundary fraction obtained after extrapolation to infinite time (intercept of lines in A with the abscissa). This represents an integral distribution $G(s)$. The data from boundary division #8 are highlighted in magenta.

6.1.3   Application to Mixtures: Diagnostic of Heterogeneity

Unfortunately the theoretical basis for the extrapolation does not generally hold in the presence of mixtures, and the results of the van Holde–Weischet method in this case are only of a qualitative nature. This can be seen by considering a mixture of two species with sedimentation and diffusion coefficients s_1 and D_1, and s_2 and D_2, respectively. A boundary division of the normalized sum of two terms of Eq. (6.2) is

$$
\begin{aligned}
\frac{i}{N} = {} & \frac{e^{-2\omega^2 s_1 t}}{e^{-2\omega^2 s_1 t} + e^{-2\omega^2 s_2 t}} \frac{1}{2} \left[1 - \Phi \left(\frac{m \left(\omega^2 s_1 t - \log \left(R_{i,t}/m \right) \right)}{2\sqrt{D_1 t}} \right) \right] \\
& + \frac{e^{-2\omega^2 s_2 t}}{e^{-2\omega^2 s_1 t} + e^{-2\omega^2 s_2 t}} \frac{1}{2} \left[1 - \Phi \left(\frac{m \left(\omega^2 s_2 t - \log \left(R_{i,t}/m \right) \right)}{2\sqrt{D_2 t}} \right) \right] .
\end{aligned}
\tag{6.5}
$$

Neglecting problems related to the scaling factors of each term of the division arising from differences in the radial dilution of the two boundaries, after rearrangement we have

$$2 - \frac{4i}{N} \approx \Phi\left((s_1 - s_{i,t}^*)\frac{m\omega^2}{2\sqrt{D_1}}\sqrt{t}\right) + \Phi\left((s_2 - s_{i,t}^*)\frac{m\omega^2}{2\sqrt{D_2}}\sqrt{t}\right). \tag{6.6}$$

Unfortunately, the time dependence of $s_{i,t}^*$ cannot be solved further in a closed form due to the problem that the error function is not linear in its arguments, i.e., $\Phi(A + B) \neq \Phi(A) + \Phi(B)$ and $\Phi^{-1}(\Phi(A) + \Phi(B)) \neq A + B$, such that the operational application of the van Holde–Weischet extrapolation scheme will produce ill-defined values. Solely for fractions that have significant contributions from only one species does Eq. (6.6) approximately conform to the theoretical framework Eqs. (6.3) and (6.4) underpinning the van Holde–Weischet extrapolation. This is illustrated for the two-species case in Fig. 6.2: Good approximations of the true s-values are obtained only in the boundary fractions from the trailing part of the slow and the leading part of the fast boundary. The characteristic diagonal transitions in between have not been subject to further analysis in the literature.

However, the diagonal features and the spread of the extrapolated s_i-values do still contain valuable information exactly due to the fact that they do not conform to the ideal single-species theory. In fact, the assessment of the homogeneity of a sample has met considerable interest over time, for example, in the 1950s in the work of Gosting [175], as well as the extensive work of Baldwin on unraveling diffusion and concentration-dependent sedimentation in polydisperse systems [111, 112, 114, 180, 181]. In this context, the combination of the approach of boundary divisions and extrapolations to $1/\sqrt{t} \to 0$ by van Holde and Weischet has created a very effective graphical diagnostic tool to gain information about homogeneity or heterogeneity of a sedimentation boundary [117], without significant numerical effort and without data fitting.[8]

When applied as a tool to determine sedimentation coefficient distributions of polydisperse systems in the presence of diffusion, due to the breakdown of its theoretical foundation in this case, it is of limited accuracy and resolving power [57]. This is illustrated in Fig. 6.3 by the application of the van Holde–Weischet method to the mixture of a 5.6 S and a 6.5 S species previously taken as a model system for $c(s)$ in Fig. 5.6. Where the direct boundary modeling approach resolved the two species, we find only a diagonal line at intermediate values in the van Holde–Weischet $G(s)$ distribution. Even in the best possible scenario — the application to non-diffusing species — the accuracy of $G(s)$ is limited due to the problems

---

[8]The extension of the diagnostic interpretation of the result of the van Holde–Weischet algorithm was proposed later by Demeler, Saber, and Hansen [182]. Briefly, reaction boundaries are suggested to generally produce positive slopes in the resulting $G(s)$ estimate, similar to heterogeneous mixtures, whereas concentration-dependent repulsive non-ideality can be the cause of negative slopes in $G(s)$ [182]. However, it is possible that these effects compensate (see green trace in Fig. 6.3) or produce more complex curves, dependent also on scan selection.

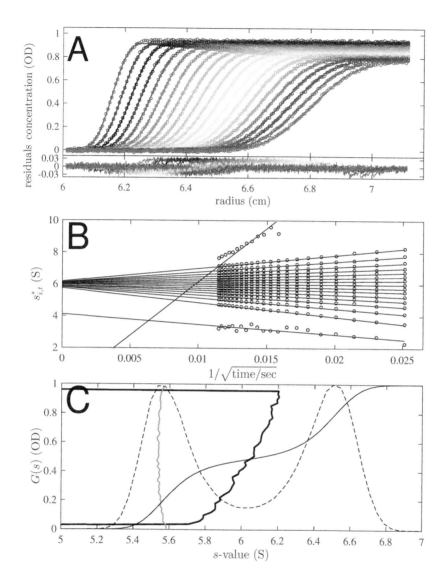

Figure 6.3 Van Holde–Weischet analysis of the two-species sedimentation process depicted in Fig. 5.6 (sedimentation of a 90 kDa, 5.6 S species and a 115 kDa, 6.5 S species at equal signal concentrations sedimenting at 50,000 rpm). *Panel A*: Sedimentation data (circles) truncated to the subset of scans 5–25 for which van Holde–Weischet analysis is applicable based on the presence of suitable plateaus. The solid lines show the best-fit radial positions of 48 boundary divisions (the maximum number ensuring each division always contains at least one data point) after extrapolation in *B*. *Panel B*: Boundary positions in the s^*-scale determined from the sedimentation boundaries (circles) and best-fit linear fit according to Eq. (6.4). For clarity, only every 3rd fraction is shown. *Panel C*: Extrapolated s_i-values constituting the $G(s)$ estimate of the van Holde–Weischet method (blue solid line). For comparison, the scaled $c(s)$ analysis from Fig. 5.6 using maximum entropy regularization at $P = 0.68$ is shown (black dashed line) and the corresponding $G(s)$ distribution after integration of $c(s)$ (black solid line). Also shown is the result of the van Holde–Weischet analysis applied to the same system at concentrations of 5 mg/ml for each species with repulsive non-ideality at k_s=0.01 ml/mg.

in boundary divisions outlined above, and the distribution problem can be solved more directly and comprehensively with ls-$g^*(s)$.

> For theoretical interest, an implementation of the van Holde–Weischet method is available in SEDFIT and can be carried out with the function Options ▷ Size Distribution Options ▷ van Holde-Weischet transform [57]. The implementation allows for a user-defined number of boundary divisions, with the radial positions determined as the average radius of data points within the limits of each fraction. This avoids the need for data smoothing. The best-fit boundaries are back-projected as step functions into the original data space for comparison against the raw data and to determine the rmsd of the model. For a good boundary representation, a high number of fractions is desirable; if too high, SEDFIT will reduce the user-defined value to the maximum possible so that at least one data point is in each division for each scan.

6.2 THE TIME-DERIVATIVE APPROACH TO $g^*(s)$ BY STAFFORD

The time-derivative approach to $g^*(s)$ by Stafford [150] is also referred to as the "dc/dt-method." Like the first step of the van Holde–Weischet procedure, it is based on the equation of motion of ideally sedimenting non-diffusing particles, which underlies the familiar coordinate transformation from r to an apparent sedimentation coefficient s^*, Eq. (4.5).[9] However, the goal of the dc/dt method is to provide an estimate of a differential sedimentation coefficient distribution, and there is no attempt to dissect contributions of diffusion from those of sedimentation. As such, it will be even more closely related to the ls-$g^*(s)$ approach introduced above.

6.2.1 Analytical Relationships

In the study of sedimentation of distributions of non-diffusing particles in Chapter 4 we derived how the sedimentation coefficient distribution $g^*(s)$ relates to the evolution of the boundary shape, specifically its spatial derivative da'/dr as expressed in the Bridgman equation (4.7) in Section 4.2, and the temporal derivative da'/dt Eq. (4.10) in Section 4.3. This provides a theoretical framework for the dc/dt method.

To understand the motivation for this method it is helpful to consider the experimental data acquisition. As described in Part I, experimental SV data $a(r,t)$ are usually subject to systematic noise, which can be decomposed into time-invariant (TI) and radial-invariant (RI) contributions. We recall that a' in the relationships of spatial and temporal derivative Eqs. (4.7) and (4.10) denotes data free of such systematic noise, although the radial derivative is naturally tolerant to RI contributions and the temporal derivative is invariant with TI contributions. The latter was exploited first by Cohen and colleagues in 1971 [184], who developed a digital absorbance scanner "MaD" that was highly sensitive, with a precision of 10^{-5} absorbance units, but revealed much larger radial-dependent offsets caused

---

[9]For this reason, the distribution was alternatively referred to as $g(s^*)$ in [183].

by unavoidable optical imperfections: this problem could be eliminated by considering the time-derivative, as approximated by time difference curves [184]. This situation is similar in the Rayleigh interference optical data of the Optima XL-I, which typically have a precision at least an order of magnitude better than the unavoidable radial-dependent fluctuations from offsets reflected in TI noise. Therefore, to fully exploit the sensitivity of optical systems, at the time, it was of interest to develop a method to determine $g^*(s)$ based on the time-derivative, rather than the established and theoretically more straightforward spatial derivative. RI noise from integral fringe shift and jitter was to be dealt with on an *ad hoc* basis in a "pre-processing" stage and was then considered absent.

The relationship between the time-derivative of the boundary da'/dt and the apparent sedimentation coefficient distribution $g^*(s)$ was already derived in Section 4.3 on p. 84. This completely describes the background of the dc/dt method. As previously noted, a key problem in the time-derivative of the boundary is the contribution from radial dilution as apparent from the two terms in the temporal analogue of the Bridgman equation Eq. (4.11) on p. 85. Even though Eq. (4.11) would be the most straightforward approach to calculate $g^*(s)$, it was not known at the time. After initially considering radial dilution effects negligible [185], Stafford later presented an approximate correction for radial dilution based on the iterative solution of the implicit integral equation Eq. (4.10) [150]. Briefly, inserting the abbreviation $\hat{g}^*(s^*) = g^*(s^*)e^{-2\omega^2 s^* t}$ in Eq. (4.10), we arrive at [150]

$$\hat{g}^*(s^*) = \frac{-t}{s^*} \left[\frac{da'(r(s^*), t)}{dt} + 2\omega^2 \int_0^{s^*} \hat{g}^*(s')s'ds' \right] . \tag{6.7}$$

The proposed scheme starts by neglecting the radial dilution term, to generate an approximation of $\hat{g}^*(s^*)$ to be used to estimate the dilution term in the following iteration, and so on [150]. After convergence of the procedure, Stafford proposed that the differences between corrected and uncorrected distributions be attributed to a correction of the time-derivative of the boundary

$$g^*(s) = \frac{t}{s^*} \times \frac{r^2}{m^2} \times \left(\frac{-da'(r,t)}{dt} \right)_{\text{corr}} , \tag{6.8}$$

in order to postulate a formula analogous to the Bridgman equation (4.7) for radial derivatives [150,183]. Unfortunately, no solution was provided that would allow one to determine the necessary quantity $(da'/dt)_{\text{corr}}$ from actual AUC data (other than its reconstruction after having first determined $g^*(s^*)$ iteratively, at which point $(da'/dt)_{\text{corr}}$ is rendered superfluous). In retrospect, comparing Eq. (6.8) with the explicit expression from the temporal analogue of the Bridgman equation in Eq. (4.11) on p. 85, it now becomes clear that the correction should have been [151]

$$\left(\frac{-da'(r,t)}{dt} \right)_{\text{corr}} = \frac{-da'(r,t)}{dt} + \frac{2}{r^{*2}} \int_0^{r'=r^*(s^*)} \frac{da'(r',t)}{dt} r'dr' . \tag{6.9}$$

The correction is the radially weighted accumulated time-derivative contributions from all smaller species, which can be readily calculated.[10]

6.2.2 Numerical Approximation of the Time-Derivative

In the configuration of commercial modern analytical ultracentrifuge detection systems, radial scans are recorded sequentially at discrete time-points, as opposed to, for example, fixed-radius detectors that continuously record [96]. This makes it necessary to approximate the time-derivative da'/dt by scan differences $\Delta a'/\Delta t$. Stafford describes a recipe for calculating an averaged $\Delta a'/\Delta t$ across a number of $2N$ scans, where time differences are transformed from the radius coordinate to the s^* coordinate and then averaged at constant s^*

$$\left\langle \frac{a_{N+i} - a_i}{t_{N+i} - t_i} \right\rangle \approx \frac{da'}{dt} \tag{6.10}$$

[183].[11] By design, this step renders the method invariant with respect to TI noise, at a cost of amplification of statistical errors [46].[12]

---

[10]For $g^*(s)$ distributions calculated on the basis of the time-derivative da'/dt, it is interesting to verify its relationship with the transport method. From the definition of s_w Eq. (2.45) on p. 52 we can draw the time-differential into the integral as

$$s_w =: \lim_{t \to 0} \left[\frac{1}{a(r',t)\omega^2 r'^2} \left(-\int_m^{r'} \frac{da'(r,t)}{dt} r\, dr \right) \right],$$

and express the time-derivative with the help of $g^*(s)$ as in (4.10) on p. 84, leading to

$$s_w =: \lim_{t \to 0} \left[\frac{1}{a(r',t)\omega^2 r'^2} \int_m^{r'} \left\{ g^*(s^*) \frac{s^* e^{-2\omega^2 s^* t}}{t} + 2\omega^2 \int_0^{s^*} g^*(s')s' e^{-2\omega^2 s' t} ds' \right\} r\, dr \right].$$

On the basis of Eq. (4.5), we can rewrite the differential $r\,dr$ as $m^2\omega^2 t e^{2s^*\omega^2 t} ds^*$ with new integration limits from 0 to s^*_{max} (corresponding to a value where $g^*(s^* > s^*_{max}) = 0$). Abbreviation of the integral in the second term as a function $h(s^*)$ leads, after cancelation of terms, to

$$s_w =: \lim_{t \to 0} \left[\frac{m^2}{a(r',t)r'^2} \left\{ \int_0^{s^*_{max}} g^*(s^*)s^* ds^* + 2\omega^2 t \int_0^{s^*_{max}} h(s^*) e^{2s^*\omega^2 t} ds^* \right\} \right].$$

In the limit $t \to 0$ the second term vanishes, and due to the square dilution law, the factor $m^2/(a(r)r^2)$ equals $1/a(m)$, i.e., a normalization to the total loading signal. As a consequence, we arrive at the expected relationship between s_w and the normalized weighted integral of $g^*(s)$. Unfortunately, the same will not hold true for $\tilde{g}^*(s, \Delta t)$.

[11]A slightly different approach resulting in smaller statistical errors was described by Philo [153], whereby $\hat{g}^*(s^*) = g^*(s^*)e^{-2\omega^2 s^* t}$ is evaluated for each scan difference first, and the resulting $\hat{g}^*(s)$ distributions are averaged.

[12]The error amplification has an N^{-3} dependence. Unfortunately, pair-wise differencing between scans that are half the total time difference apart has a maximum error amplification with regard

In principle, from the point of view of achieving a good signal/noise ratio and precise s-values, it would be desirable to use a large time interval, allowing for a significant boundary migration between the scans to be subtracted, and providing a large number of scans to average. On the other hand, this would be detrimental to the approximation of the time-derivative. As a compromise, for the choice of the total time interval $\Delta t = t_N - t_1$, Stafford has recommended an empirical "rule of thumb"

$$\Delta t = \frac{160\, t}{(\text{rpm}/1000)\sqrt{M/kDa}} \tag{6.11}$$

(where M is the mass of the sedimenting particle) or half that value when diffusion coefficients are to be estimated [153].[13]

The quality of the approximation Eq. (6.10) is critical for the performance of the dc/dt method, and therefore a more rigorous analysis is warranted. Because the theory of dc/dt is predicated on the analysis of migrating step functions of non-diffusing species, it is straightforward to sketch the effect of the approximation $dc/dt \approx \Delta c/\Delta t$ by theoretically examining the fate of boundaries from a single non-diffusing species at unit concentration [121,151]. In this case the boundaries are the familiar migrating step of Eq. (2.10), and the sedimentation coefficient distribution is simply $g_1^*(s) = \delta(s - s_1)$ if we denote the s-value of the species as s_1. In this case, their time difference over the time interval Δt from t_1 to t_2 is

$$\frac{\Delta a'(r,t)}{\Delta t} = \frac{1}{\Delta t} \begin{cases} 0 & \text{for } r < me^{s_1\omega^2 t_1} \\ -e^{s_1\omega^2 t_1} & \text{for } me^{s_1\omega^2 t_1} < r < me^{s_1\omega^2 t_2} \\ e^{s_1\omega^2 t_2} - e^{s_1\omega^2 t_1} & \text{for } r > me^{s_1\omega^2 t_2} \end{cases} \tag{6.12}$$

Inserted into Eq. (4.11) in place of the time-derivative, this leads to an apparent distribution that we may designate $\tilde{g}_1^*(s, s_1, \Delta t)$ [151]

$$\tilde{g}_1^*(s, s_1, \Delta t) = \frac{t}{s^*\Delta t} \begin{cases} 0 & \text{for } s^* < s_1 t_1/t \\ 1 & \text{for } s_1 t_1/t < s^* < s_1 t_2/t \\ 0 & \text{for } s^* > s_1 t_2/t \end{cases} \tag{6.13}$$

This means that, as a result of the replacement of the time-derivative by the time difference, the true distribution $g_1^*(s) = \delta(s - s_1)$ of a single species will turn into an apparent distribution that is shaped like a hyperbola segment of width $\Delta s = s_1\Delta t/t$. [121,151]. This is illustrated in Fig. 6.4.

---

to the boundary velocity; a minimum error amplification with a pair-wise differencing scheme is achieved when scans are subtracted that are one-third of the total time difference apart [46]. The best strategy would correspond to a difference scheme using the average scan as a reference, which is equivalent to algebraic TI noise elimination method [46] (see Part I, Chapter 4).

[13]The appearance of a molar mass in this formula may seem confusing at this point, but the dc/dt method was developed with diffusion-broadened boundaries in mind, which translates into to a broadening of the apparent sedimentation coefficient distribution, subject to interpretation in a second stage.

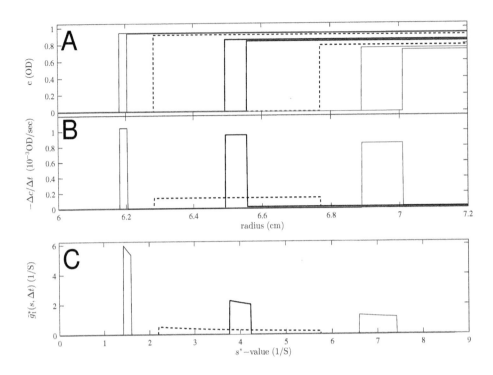

Figure 6.4 Relationship between boundary profiles of non-diffusing species (*Panel A*), their time differences (*Panel B*), and the response function $\tilde{g}_1^*(s, s_1, \Delta t)$ (*Panel C*) in the dc/dt-method. The boundaries are for a single 4 S species with time intervals and boundary positions corresponding to Fig. 4.3 on page 87, with the narrow time interval from 7200 to 8100 sec comprising 4 scans (number 24 to 27) as in Fig. 4.3A shown as black solid lines, and the large time interval from 4200 to 11,100 sec comprising scans 14 to 37 in Fig. 4.3B shown as black dotted lines. For the short time interval, for comparison, analogous data are shown for a single non-diffusing species at 1.5 S (blue) and 7 S (magenta).

We can now go beyond the single-species model to entire distributions of species $g^*(s)$, which are nothing more than superpositions of single species. Therefore, we may assemble the resulting $g^*(s)$ distribution directly from superpositions of the response function of a single species. This idea can be phrased as the identity $g^*(s) = \int g^*(s')\delta(s - s')ds'$, and therefore $g^*(s) = \int g^*(s')g_1^*(s')ds'$. All signals being additive, the apparent distribution $\tilde{g}^*(s, \Delta t)$ obtained after using $dc/dt \approx \Delta c/\Delta t$ in the dc/dt method can then be understood as a convolution of the true distribution with the response function $\tilde{g}_1^*(s, s_1, \Delta t)$ consisting of hyperbolic segments [121, 151]:

$$\tilde{g}^*(s, \Delta t) = \int g^*(s')\tilde{g}_1^*(s, s', \Delta t)ds' . \tag{6.14}$$

The convolution function has a relative width $\Delta s/s$ equal to the fractional time $\Delta t/t$ used for the approximation of dc/dt. For example, if the data for analysis is comprised of scans during which the boundary migrates 10% farther from the

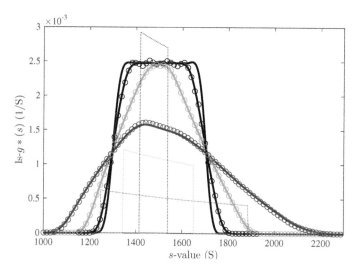

Figure 6.5 $g^*(s)$ distribution of the simulated sedimentation data of the broad uniform distribution from $1300 - 1700$ S from Fig. 4.1C/D on page 78. In black is shown the ls-$g^*(s)$ distribution fit to data between 450 and 1,020 sec. It exhibits smooth but relatively steep gradients at the edges of the distribution due to a small degree of diffusion broadening. The results from $g(s^*)$ in the dc/dt method applied to the same time interval is shown as a bold magenta solid line. The restriction of the time range to 600–870 sec is shown as a bold cyan solid line, and a further restriction to 690 – 780 sec as a blue bold solid line. $g(s^*)$ was calculated using the program DCDT+ kindly provided by John Philo. The response functions $\tilde{g}_1^*(s, s', \Delta t)$ for the middle of the time intervals is shown as a thin dotted line. To demonstrate the convolution Eq. (6.14), circles show $\tilde{g}^*(s, \Delta t)$ calculated with the response functions convoluting the ls-$g^*(s)$ distribution (taken for the true distribution). Reproduced from [151].

meniscus, then the convolution in s will span $\sim 10\%$ of the mean s-value of the boundary. Unfortunately, because the convolution is not simply a box-average but a segment of a hyperbola, the resulting shape of $\tilde{g}^*(s, \Delta t)$ will be shifted toward smaller s-values, an effect exacerbated at smaller s-values [121]. These theoretical expectations match very well with empirical observations [121,153].[14] Examples are Figs. 6.5 with a hypothetical rectangular large species distribution that shows the distortions most clearly, Fig. 6.6 with simulated data for a small protein showing distortions in the diffusion envelope, and Fig. 6.7 with the application to experimental BSA data.

With the danger of the interpretation going astray from the information content of the SV data, as always, it is prudent to back-project the resulting distribution into the original data space, to examine what boundary features are captured in this interpretation. If the $g^*(s)$ distribution is correct, then, by design, the g^*-values describe the amplitudes of a step function of a non-diffusing species, and their superposition must describe the experimental sedimentation boundaries. This is equivalent to substituting $g^*(s)$ for ls-$g^*(s)$ in Eq. (4.1). Best-fit TI and RI noise

---

[14]Exact results may depend on details of the implementation and possibly additional numerical operations, such as averaging, smoothing, or "zeroing."

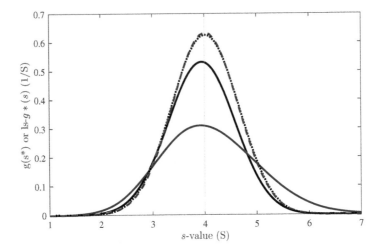

Figure 6.6 $g^*(s)$ distribution of the simulated sedimentation data of the 4 S, 80 kDa species from Fig. 4.3 on page 87, using the same the narrow time interval from 7200 to 8100 sec comprising 4 scans (number 24 to 27) in Fig. 4.3A (black solid line), and the large time interval from 4200 to 11,100 sec comprising scans 14 to 37 in Fig. 4.3B (magenta solid line). ($g(s^*)$ was calculated using the program dcdt kindly provided by Walter Stafford.) The response function $\tilde{g}_1^*(s, s_1, \Delta t)$ for these time intervals are shown in Fig. 6.4C. For comparison, the results from the ls-$g^*(s)$ distribution analysis of the same data are shown as black dotted and magenta dashed lines.

can then be calculated while keeping $g^*(s)$ fixed, and residuals of the fit can be inspected.[15] Unfortunately, this quality control step has traditionally not been part of the dc/dt-method. In fact, due to the broadening artifact from finite Δt, traces may not fit the raw data very well (as shown, e.g., Fig. 3 in [107]).

The SEDFIT function Options ▷ Size Distribution Options ▷ transform g(s*) (as ASCII) to direct boundary model can accomplish this task, based on user-provided $g(s^*)$ distributions that have been calculated by the software dcdt, SEDANAL, or DCDT+.

6.2.3 Applications of dc/dt-derived $g^*(s)$ or $g(s^*)$

The basic necessity to approximate a snapshot of the time-derivative to avoid excessive broadening in the dc/dt method requires the experimenter to disregard the vast majority of the sedimentation data. For example, if broadening by $\Delta s/s$ of less than 10% is deemed acceptable, then ~90% of the boundary data has to be discarded and left uninterpreted (or more, considering the scans from later time points). For absorbance data, which often have comparatively large time intervals between subsequent scans, only a few scans may remain in the analysis if tight accuracy limits are required. Further, all species to be reflected in the distribution have

---

[15]Conceivably it could offer a rational guide for selection of Δt in a "model-free" way.

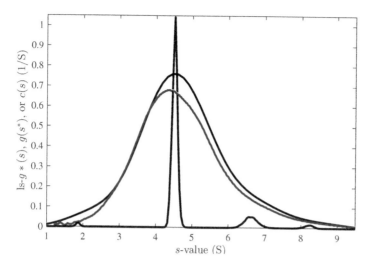

Figure 6.7 Analysis of the BSA data of Fig. 5.1 on p. 93 by dc/dt (magenta), ls-$g^*(s)$ (blue) and $c(s)$ (black). For the dc/dt analysis only scans in the time range 4168 to 5191 sec were included in the analysis by dcdt software (kindly provided by Walter Stafford), which adheres to the rule of thumb Eq. (6.13) for species including the ~8 S trimer. The ls-$g^*(s)$ model was applied to scans in the range 2480 to 6698 sec, and the $c(s)$ model to all scans comprising 190 to 7844 sec.

to show migrating boundaries within the radially observable window at the same time, significantly truncating the size range of species that can be studied concurrently.[16] Finally, contrary to its theoretical foundation on non-diffusing species, the artificial Δt convolution is least noticeable for conditions where broad boundaries are generated, such as diffusion boundaries of moderately large proteins at not too high rotor speeds, sacrificing hydrodynamic resolution at higher fields [121].

From a modern perspective — with the ability to freely fit explicit distribution models directly to all available raw data without any of these constraints, for example using ls-$g^*(s)$ or $c(s)$ — this seems very limiting (Fig. 6.7). However, it must be kept in mind that the development of the dc/dt method historically coincided with the development of digitized interference detection systems (including the commercial Optima XL-I) [187, 188], which produced precise fringe shift data at a high rate but with overwhelming TI noise contributions. Following Cohen and colleagues [184], the dc/dt method with its scan differences addressed the TI noise

---

[16]This rules out applications to particle distributions such as shown in Fig. 4.2; other examples can be found in [121]. An approach to address this limitation by combination of $g(s^*)$ traces obtained at sequentially higher rotor speeds applied in the same run, termed "wide distribution analysis" is described in [99] and applied in [186]. Since all equations describing sedimentation in dc/dt depend on $\omega^2 t$, scans at each rotor speed step can be analyzed in the conventional way and incur the conventional convolution artifact. However, multiple time steps at different speeds invalidate assumptions in the standard analysis of diffusional spread in $g(s^*)$ [22]. In comparison, ls-$g^*(s)$ is naturally compatible with time-varying centrifugal fields [22], but also inherently suitable for analyses of samples with a large s-range where particles transition through the observed radial window at different times (e.g., Fig. 4.2).

problem with the methods available in the early days of routine computer analysis. As a distribution method it does not require the specification of the number of sedimenting species, and offers two significant improvements over the van Holde–Weischet analysis: First, it is compatible, in principle, with the transport method to determine s_w and therefore offers an opportunity to characterize interacting systems.[17] Second, for sufficiently homogeneous samples it provides an opportunity for determining approximate diffusion coefficients and buoyant molar masses in a second-stage analysis. Thus, the interpretation of $g^*(s)$ of diffusion-broadened boundaries in terms of Gaussians arising from single-species sedimentation and diffusion coefficients was one of the most common applications of the dc/dt method. In this context, Stafford switches to the terminology $g(s^*)$ rather than $g^*(s)$ to emphasize that s ceases to be a true sedimentation coefficient of a sedimenting species and becomes merely a transform of the radial coordinate [183].

The analysis of diffusion coefficients from $g(s^*)$ proceeds in the same way as outlined for the ls-$g^*(s)$ distribution in Section 4.4, with a Gaussian model for the apparent sedimentation coefficient distribution Eq. (4.13) based on the first term of the Fujita–MacCosham approximate solution of the Lamm equation Eq. (2.31). This approach has the same caveats with regard to the lack of resolution of polydispersity within the diffusion-broadened boundary when using small time intervals of data as illustrated in Fig. 4.6 (p. 90). Unfortunately, extension of the time interval of scans to address this problem is not possible in dc/dt. With regard to the accuracy of the parameters for homogeneous samples, fortuitously, errors in the Gaussian fit from the limited accuracy of the first term Fujita–MacCosham approximate Lamm equation solution, which tend to underestimate diffusion coefficients and over-estimate molar masses [10, 121] (see Fig. 4.3 on p. 87), can under some conditions be partially compensated by the artificial broadening of $g(s^*)$ from the $dc/dt \approx \Delta c/\Delta t$ approximation, which itself tends to underestimate the molar masses [121, 153]. Complicating matters is the increasing skew toward underestimates of s-values for slowly sedimenting species.

A useful property of the second stage $g^*(s)$ analysis is the opportunity to truncate the fitted data range to a certain range of s^*-values, which can sometimes help to decrease the impact of sample polydispersity in $g^*(s)$. In direct boundary modeling, an equivalent approach exists and is termed "partial boundary modeling" [38] (see Section 8.1.4.1). A detailed comparison of the performance of $g^*(s)$-based analysis with modern direct boundary modeling techniques can be found in [38].

Several improvements were developed by Philo [152, 153], including a direct fit of the average time difference data, and better approximations of the Lamm equation solution [153]. Further, taking advantage of increased computational power available, Philo described the fit of the experimental $\tilde{g}^*(s)$ data with theoretical $\tilde{g}(s^*)$ traces of single-species Lamm equation solutions that have undergone the same scan selection and computational steps as the experimental scans [152]. While

---

[17]However, the skewed broadening of $g^*(s)$ from dc/dt will cause s_w to be underestimates, increasingly so at smaller s-values and larger $\Delta t/t$ [107].

the latter eliminates many of the sources of error from the approximations of the original dc/dt method, it is still subject to the same destruction of information that occurs when compressing the information from many experimental scans into a single $g^*(s)$ trace, which would be apparent when considering the back-projection of the $g^*(s)$ into its associated boundary model for comparison with experimental data. It also requires empirical removal of RI noise [46]. Finally, it is subject to the lack of resolution and inherent range limitation of the dc/dt discussed above.[18]

6.3   LIMITING RELATIONSHIP BETWEEN $G(s)$ AND $g^*(s)$

Problems and limitations in the practical application of van Holde–Weischet analysis and the dc/dt method aside, it is theoretically interesting to examine the relationship between the two [57]. The connection can be established best on the basis of the step function model of the ls-$g^*(s)$ distribution. Let us divide up the total number of scans into groups i with moderate time range Δt, centered at t_i. For each group, we use the ls-$g^*(s)$ method to determine $g^*(s, t_i)$. Now, we integrate each $g^*(s, t_i)$ to arrive at N equal area fractions. In each fraction, we take the weighted average s-value: Since the area fractions of $g^*(s, t_i)$ correspond to boundary fractions, the former will be excellent estimates of the $s_{t,i}^*$-values required in the van Holde–Weischet method for extrapolation to infinite time. Again, we extrapolate the best-fit straight lines of $s_{t,i}^*(\sqrt{t_i})$ back to $\sqrt{t_i} \to 0$, corresponding to the limit $t \to \infty$, to arrive at $G(s)$ [57].[19]

Conceptually, there are two attractive features in this strategy for determining $G(s)$: 1) It is apparent that the area fractions of ls-$g^*(s)$ are recombining the original steps of ls-$g^*(s)$ (which may be many more than possible to use in the van Holde–Weischet method) to a suitable number of steps of equal height. In contrast to the original van Holde–Weischet method, however, the radial dilution of species sedimenting at different positions is properly taken into account, since the division into equal fractions is made in terms of equivalent loading concentration (which is the scale of ls-$g^*(s)$). 2) The initial ls-$g^*(s)$ analysis can account for TI and RI noise, in this way making the van Holde–Weischet procedure applicable to data from the interference optical detection system, or any other data that has time-invariant noise contributions.

---

[18]Stafford and Sherwood [62] have described a further logical step in the improvement of the analysis, which is to abandon the attempt to determine a $g^*(s)$ distribution completely, and instead fit directly all the $\Delta c/\Delta t$ traces to a discrete Lamm equation model. This removes the need for averaging $\Delta c/\Delta t$ and the associated information loss for large time intervals. Lacking consideration of polydispersity, conceptually, this strategy is closely related to a discrete species model in direct boundary analysis: Unfortunately, these usually do not describe experimental data very well due to almost ubiquitous sample imperfections and polydispersity, accounting for which requires distribution methods. Further, the lack of rigorous account for RI noise and unnecessary noise amplification are remaining drawbacks of the $\Delta c/\Delta t$ approach [62].

[19]In 1959, Baldwin [111] introduced a similar extrapolation based not on area fractions of $g^*(s)$ but on fractions of distribution values $g^*(s)/g^*(s)_{max}$.

Conversely, we can think of this approach as a way to exploit the time dependence of the $g^*(s)$ distribution: As was shown in Fig. 4.4 on p. 88, when calculating $g^*(s)$ for diffusing systems, the sharpness of the peaks increases with time. The approach of calculating area fractions and extrapolating their average $s_{t,i}^*$-values to infinite time is a simple means to enhance and exploit this dependence to arrive at a diffusion-free sedimentation coefficient distribution [57].[20]

It must be kept in mind that these are theoretical considerations that highlight connections between historic approaches, but that a much more rigorous and practically feasible approach for diffusion-deconvoluted sedimentation coefficient distributions is, of course, the model of the sedimentation boundary directly with the $c(s)$ distribution or analogous, as described in Chapter 5.

To facilitate the theoretical comparison of methodologies, and as a complementary method to the van Holde–Weischet implementation in SEDFIT, the function Options ▷ Size Distribution Options ▷ extrapolate area fractions of ls-g*(s) vs 1/sqrt(time) will carry out the described extrapolation of area fractions of ls-$g^*(s)$ *vs.* $1\sqrt{t}$. However, this is not the recommended approach for determining diffusion deconvoluted sedimentation coefficient distributions.

6.4 DATA TRANSFORMATIONS, MODEL-FREE, AND ALGORITHMICALLY DEFINED APPROACHES

Modern boundary modeling techniques follow principles outlined in Chapter 1. These carry the burden for the investigator to select a model appropriate for the system under study, and to achieve a fit ideally with residuals close to within the random noise contribution of the data acquisition, or at least without systematic mismatch of sedimentation boundary features (see Part I, Section 4.3.5). For routine SV analysis this represented a change in paradigm that was not possible before computational resources were abundant. By contrast, the van Holde–Weischet and time-derivative approaches have been characterized as being "model-free" [179,189], and that one should regard them merely as unbiased "data transforms" — and as such, by definition, they should always be understood as a correct reflection of the properties of the sedimentation data. This is not a scientific viewpoint.

Strictly, the only model-free method to obtain insight from experimental data is mystical. Each physics-based method for the quantitative data analysis in SV relies on fundamental equations that embody a model of the sedimentation process in the gravitational field. Specifically, both the van Holde–Weischet and the dc/dt

---

[20]The extrapolation of $g^*(s)$ to infinite time has been suggested before by others with various success [24,111,175,185]. Stafford [185] describes difficulties in the choice of the appropriate extrapolation procedure, including the problem of extrapolations to negative distribution values. One of the key differences between the earlier [185] and the present approach is that in the procedure described here, the extrapolation of g*(s) is carried out not at constant s, but at constant area of $g^*(s)$, which is similar to the approach taken by Baldwin [111].

method are based on the expectation that non-diffusing particles form a boundary that propagates as $r^{(p)}(t) = me^{\omega^2 st}$. This assumption enables us to correlate the particle's s-value with a certain radius at a specific time, and this constitutes a model that can be true or false depending on the given experiment.[21] In the van Holde–Weischet method, the second stage proceeds then by applying an idea that the particles are diffusing, but are monodisperse. This, again, is an explicit model that can be found being stated very clearly in the equations on which the method is based. This will be true for single ideal species, but false for polydisperse mixtures. The dc/dt method, conversely, is based on a model that there are distributions of particles with different s-values, each sedimenting ideally but not diffusing. There are features in the results from these methods that have been taken as flags to indicate these assumptions do not hold for a particular data set. In some cases, further assumptions can be made to model the results.

A temptation to regard these methods as being "model-free" may arise from their original design not examining residuals of the model compared to the experimental data, although now this could be easily done.[22] Obviously, the lack of a display of residuals to the raw data does not mean there was no model at play interpreting the raw data — it just means part of the computation is missing and residuals are not inspected. This is a drawback, since residuals can reveal sources of additional information that were not captured in the model yet, but could potentially be exploited in a refined model.

The term "model-free" could be misunderstood as "parameter-free": neither the van Holde–Weischet nor the dc/dt approach have obvious physically meaningful fitting parameters that could be adjusted (if one considers the straight line extrapolations in the van Holde–Weischet method as merely a procedural step). This is due to the fact that the models are very simple, as was necessary at the time.

This leads to the second term often used in the context of this discussion, which is that of a "data transform." The suggestion is that in applying these methods we are not really curve-fitting the raw data, but rather presenting it differently, within a different coordinate system: For example, the transformation of the r-axis into an s^*-axis is doing just that (using a certain model of propagation), and arrives in a space that is perhaps intuitively easier to interpret. The principal statistical drawbacks of this approach in general have long been recognized and are well known, for example, in the context of binding analyses with Scatchard plots [71, 190]. One particular additional problem with this approach in SV analysis is that the trans-

---

[21]For example, it will be false even for non-diffusing particles for the sedimentation at very high concentrations, where repulsive steric or electrostatic forces are present. Also, this assumption is false for suspensions of particles with attractive forces.

[22]Residuals for the van Holde–Weischet approach were proposed [57], with the representation of best-fit regressions of $s_{t,i}^*$ back in the data space as boundary fractions. For the time-derivative approach, residuals appear *via* the back-transform implemented in SEDFIT and presented first in [107].

formation into a simpler coordinate system of $g(s)$ or $G(s)$ leads into a space of lower dimensionality: This diminishes the possibility to verify that the correct interpretation is made. Further, both the van Holde–Weischet and the time-derivative method do apply further processing steps such as extrapolation and averaging, based on a physical model or mathematical assumptions, respectively, which help to reduce the two-dimensional spatio-temporal data set to a one-dimensional curve. These processing steps are central to these methods but can be sources of artifacts, such that the result may not faithfully represent the original data anymore. This clearly removes these approaches from the realm of mere data transforms. Examples for true data transformations include Laplace transforms, Fourier transforms, and singular value decomposition, which are mathematically exact representations of the same data in different parameter space, but do not introduce error or reduce dimensionality.

Lastly, because the van Holde–Weischet and the dc/dt method are, in principle, procedurally well defined,[23] and because of the lack adjustable physical parameters, one could regard the outcome simply as the result of the application of a sequence of certain numerical operations. In this view, the analysis would resemble a transfer function in system theory, and to the extent that the output result is an objective reflection of properties of the input data, the result is correct if the operations have been applied correctly. However, reproducible computational algorithms do not *per se* provide mathematically correct solutions to certain problems, or even scientifically useful results. A case in point is the 2DSA analysis developed by Brookes, Cao, and Demeler, which is aiming at diffusion-deconvoluted size-and-shape distribution analysis of SV data [191]. It is not further discussed in the present book, as the method side steps known matrix algebra, which is substituted by elaborate novel and demonstrably incorrect numerical algorithms exploiting supercomputer resources [137].[24] At best, this method may provide procedurally well-defined results, but whether they are scientifically meaningful is uncertain. Thus, procedural definition is essential but not sufficient for useful data analysis. Rather, the computational procedures must also be mathematically meaningful and relate to a physical interpretation of the process observed.

In summary, both the van Holde–Weischet and the dc/dt method are essentially simple parameter-free models of sedimentation boundary propagation. They have been extremely useful in the past, but like all models, they have their assumptions and limits of applicability and performance. As with any other model, it is essential to establish whether or not they faithfully represent the experimental data, and modern computers enable us to do this.

---

[23]Invariably, though, bias will enter the analysis in practice due to the need to select meniscus and data limits, systematic noise removal, "zeroing," etc.

[24]As shown in [137], even the case for using supercomputers to enable the necessary computations in the 2DSA method is vastly overstated, memory requirements being miscalculated and artificially inflated by a factor of 10 to 100.

CHAPTER 7

Multi-Component Distributions

D ISTINCTIVE optical properties in conjunction with enriched data sets can be exploited to determine separate sedimentation coefficient distributions of multiple components in a sedimenting mixture. This is particularly useful when studying interactions between dissimilar classes of sedimenting particles. Multi-component distribution methods have been developed based on multispectral analysis, and on spatio-temporal signal modulations.

7.1 MULTI-SIGNAL MULTI-COMPONENT SEDIMENTATION COEFFICIENT DISTRIBUTIONS

7.1.1 Basic Concept

SV experiments offer an additional data dimension that can be created by exploiting the different detection wavelengths or detection systems that are simultaneously available for recording signal profiles in the same run [35].[1] For example, in the Optima XL-A/I analytical ultracentrifuge the interference optical system can image the refractive index profile in the solution column, alternating with the absorbance system sequentially scanning the same solution column at up to three selected wavelengths.[2] In this way, one SV experiment can be observed with up to four different characteristic signals, opening the way to simultaneous spectral decomposition and hydrodynamic analysis of the sedimenting particles, termed multi-signal $c_k(s)$. Very recently, different instrumentation has been developed with enhanced

---

[1] For a more complete description of the spectral data dimension, see Part I, Section 4.4.

[2] It was claimed [192] that the acquisition of radial scan data at multiple wavelengths in the Optima XL-A/I analytical ultracentrifuge requires separate centrifuge runs. As is widely known, this is patently false: Radial scans at up to three wavelengths can be acquired concurrently, as specified in the scan details of the standard user interface offered since the introduction of the instrument in the early 1990s, quasi-simultaneously with Rayleigh optical data acquisition. This has been widely used.

multi-wavelength detection capabilities, allowing faster scans or simultaneous multi-wavelength acquisition. For example, the detector described in [36,193] was shown to allow the determination of absorbance spectra for hydrodynamically well-separated sedimenting species.

Due to the time required for data acquisition in the Optima XLA/I system, in particular for the absorbance scans, the time points for the radial scans at different signals are offset relative to each other, and a smaller number of scans can be collected at any single signal than the maximum possible if only a single detection were used. This can easily be accurately accounted for, and does not have much detrimental effect on the precision of the sedimentation analysis: The data complement each other in reporting the radial positions of the boundary, and jointly provide a set of time points similar to that of absorbance optical detection at a single wavelength. In our experience, it is usually sufficient to obtain approximately 10–20 radial scans per signal. This effect is insignificant and no reduction in standard scan number occurs when using interference optical detection in conjunction with absorbance at a single wavelength. This and other experimental considerations are discussed in more detail in Part I, Section 4.4.4.

Multi-signal detection is very attractive when studying mixtures of different ensembles of particles exhibiting different optical properties, or particles that consist of components with different optical properties. This includes samples with mixtures of proteins and nucleic acids, mixtures of proteins with different absorbance properties (from extrinsic chromophores, or different ratios of aromatic amino acids), proteins and detergents, or mixtures of other classes of chemically dissimilar macromolecules. For systems of this kind, the multi-signal analysis removes a major difficulty from AUC analysis, which is the attribution of the identity of different peaks in the sedimentation coefficient (or molar mass) distribution to different protein complexes. An important type of application is the study of mixtures of different protein components that form hetero- and mixed oligomers. Multi-signal SV provides unique opportunities to resolve and characterize many co-existing binary and ternary complexes and their assembly into larger multi-protein complexes, such as adaptor protein complexes in signal transduction [194–196], as illustrated in Part I, Section 4.4.4 and reviewed in [197,198].

To enable the spectral analysis, it is necessary to acquire a total number of signals Λ equal to or larger than the number of different (spectral) components K. In the context of multi-signal $c_k(s)$, we define a component as a set of spectrally identical particles. For example, a sample of a particular purified protein A — irrespective of its aggregation state or mixture of states — constitutes a component since the amino acid composition determines the extinction coefficients ε_λ at the wavelengths λ, and the molar mass (and glycosylation) determine the molar fringe coefficient $\varepsilon^{(IF)}$. In the following we will use the symbol ε_λ for both, since the molar fringe coefficient and the molar extinction coefficients can be treated identically. Further, we will use a subscript A to label this component. Thus, the set of $\varepsilon_{A,\lambda}$ defines the optical properties, or spectral signature, of the component A, and they are assumed to be constant and independent of the association state, which may include dimers AA, trimers AAA, and other oligomers A_n.[3]

Analogously, a different purified protein component B has the optical properties defined by $\varepsilon_{B,\lambda}$. In general, a mixture of the two components may exhibit many different species of different size, composition, and quaternary structure in solution, such as A, AA, A_n, B, BB, AB, AAB, ABA, A_nB_m, etc. It is easy to see that if we had — hypothetically — a signal λ_A that reported exclusively on A, the sedimentation coefficient distributions would show peaks for the following species with the signal amplitudes: $1 \times \varepsilon_{A,\lambda_A} \times [A]$, $2 \times \varepsilon_{A,\lambda_A} \times [AA]$, $n \times \varepsilon_{A,\lambda_A} \times [A_n]$, $1 \times \varepsilon_{A,\lambda_A} \times [AB]$, $2 \times \varepsilon_{A,\lambda_A} \times [AAB]$, $2 \times \varepsilon_{A,\lambda_A} \times [ABA]$, $n \times \varepsilon_{A,\lambda_A} \times [A_nB_m]$, etc., and similarly, a signal λ_B exclusive for B would show a sedimentation coefficient distribution with peaks corresponding to $1 \times \varepsilon_{B,\lambda_B} \times [B]$, $2 \times \varepsilon_{B,\lambda_B} \times [BB]$, $1 \times \varepsilon_{B,\lambda_B} \times [AB]$, $1 \times \varepsilon_{B,\lambda_B} \times [AAB]$, $1 \times \varepsilon_{B,\lambda_B} \times [ABA]$, $m \times \varepsilon_{B,\lambda_B} \times [A_nB_m]$, etc. From the combination of the peak locations and amplitudes of both sedimentation coefficient distributions, one could sort out the composition and identity of the different complexes that occur in solution (Fig. 7.1). Together with the complex s-values, one could deduce the stoichiometry and hydrodynamic shape.

While this gedankenexperiment clarifies the goal of what we want to achieve with multi-signal $c_k(s)$, in reality it is important to note that we usually do not have the luxury of a signal exclusively reporting on only one component, nor do we always have nicely separated peaks in the sedimentation coefficient distribution. However, this is not at all necessary: As long as the extinction coefficients $\varepsilon_{A,\lambda}$ and $\varepsilon_{B,\lambda}$ are sufficiently different at some signals λ_1 and λ_2, we can achieve a spectral deconvolution of each component's contribution to each signal, as we will discuss in more detail below. This can be achieved simultaneously with the diffusional deconvolution, in a global fit to all available data at all signals. This is the idea of the multi-signal $c_k(s)$ [35].

The mathematical implementation rests on a global model for the sedimentation data at each signal $a_\lambda(r,t)$ as a superposition of different $c(s)$ distribution for each component k:

$$a_\lambda(r,t) \cong \sum_{k=1}^{K} d\varepsilon_{k,\lambda} \int_{s_{min}}^{s_{max}} c_k(s)\chi_1(s, D_k(s), r, t)ds + b_\lambda(r) + \beta_\lambda(t) \qquad (7.1)$$

with $\lambda = 1 \ldots \Lambda$, where Λ is the total number of signals, and the radial- and time-invariant noise components being specific to each signal. The total number of signals must be equal to or larger than the total number of components, $\Lambda \geq K$. For this to work, the signals recorded must ensure that the extinction coefficients are at least linearly independent, $\det(\varepsilon_{k,\lambda}) > 0$ (in other words, not proportional to each other). A more stringent condition will be derived below. The $c_k(s)$ distribution then consists of several components k, each showing species that have the spectral

---

[3]If spectral changes occur, for example, in complexes due to hypo- or hyperchromicity, then these complexes constitute a separate spectral component in the analysis.

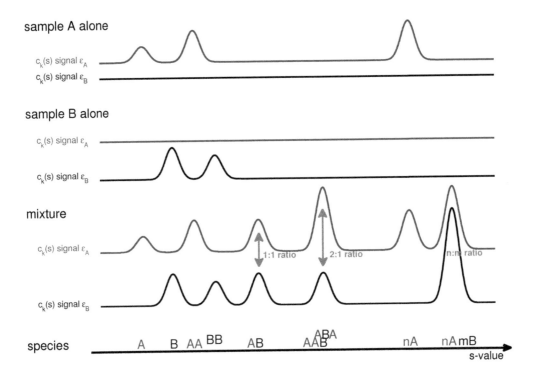

Figure 7.1 Principle of multi-signal $c_k(s)$ assuming hypothetically that there are signals λ_A and λ_B exclusively reporting on components A and B, respectively. In a sample that only contains component A, when analyzing the signal λ_A we would obtain a sedimentation coefficient distribution reflecting only A, and potentially its oligomeric species, while signal λ_B would vanish. Similar is true for a sample containing component B alone. The analysis of the mixture may reproduce the peaks of the individual components, but also additional peaks at certain s-values arising from complex formation between A and B. The ratios of the signal amplitude (peak integral) of these peaks in signal λ_A and λ_A reveal the molar ratio of the components in the complex. Jointly with the s-value, the stoichiometry can be derived (since, for example, a 1:1 complex sediments at a very different s-value than a 2:2 complex). However, different possible structural configurations of complexes exhibiting the same stoichiometry cannot be distinguished (such as ABA from AAB in the above scheme).

signature $\varepsilon_{k,\lambda}$, i.e., that have contributions migrating at the particular s-value with the same signal ratio in all signals as component k. The scaling relationship $D_k(s)$ in Eq. (7.1) may be chosen such that it reflects only species that have this spectral signature. Usually, $D_k(s)$ will rest on the scale relationship for compact particles with an average $f_{r,w}$. Since the signal increments $\varepsilon_{k,\lambda}$ in Eq. (7.1) are in molar units, the component sedimentation coefficient distribution $c_k(s)$ will, likewise, be in molar units (in contrast to the regular $c(s)$ distribution which is simply in experimental signal units).

7.1.2 Choice of Spectral Signals

The multi-signal model Eq. (7.1) is useful in two ways: First, in separate experiments with the individual components κ (where $k = K = 1$), we can determine simultaneously the extinction coefficients $\varepsilon_{\kappa,\lambda}$ and the sedimentation coefficient distribution of this component (Fig. 7.2). At this stage, all but one $\varepsilon_{\kappa,\lambda}$ will be treated as a parameter to be refined in the fit. As described in more detail in Part I, Section 4.3.2, this approach (and its extension to the segmented hybrid discrete/continuous multi-signal distribution Eq. (7.6) below) is the optimal approach for determining extinction coefficients by AUC. It simultaneously accounts for the size distribution of a component $c_\kappa(s)$ and potential impurities while determining the extinction coefficients.

As it rests on global analysis, the multi-signal $c_k(s)$ analysis is implemented in SEDPHAT. It can be invoked by the function Model▷Multi-Wavelength Discrete/Continuous Distribution Analysis. In the Global Parameters an area in the center right outlined by double lines has the entry fields for the $\varepsilon_{k,\lambda}$, with coefficients for each component in the same row, and at the same signal in the same column. Molar extinction coefficients can be entered in the respective fields. Per convention, the checkbox at each field can mark a particular parameter to be refined in the non-linear regression initiated by the FIT command.

Inspection of Eq. (7.1) shows that fitting all extinction coefficients along with the distribution would create a redundant set of unknowns. Thus, it is essential for each component to introduce a known extinction coefficient at exactly one of the signals, in order to arrive at a well-defined molar concentration scale. Practical considerations for the choice of this spectral "fixed point" can be found in Part I, Sections 4.3.2 and 4.4.4. For example, the signal increments in the refractive index-sensitive Rayleigh interference optical detection system are very similar for most proteins [108], except for small peptides and special classes of proteins [108, 199], as well as for proteins with significant posttranslational modification including glycosylation. The latter, however, leaves aromatic UV absorbance of the protein invariant. Depending on the protein under study, therefore, the refractive index

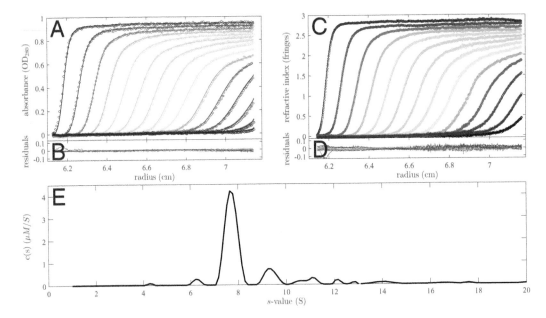

Figure 7.2 Single component multi-signal SV data of a preparation of a glycosylated viral coat protein, acquired by absorbance at 280 nm (*Panels A and B*) and Rayleigh interference detection (*Panels C and D*), and global fit with a single component $c_k(s)$ (*Panel E*). In the global fit, both the sedimentation coefficient distribution as well as the interference optical signal increment were determined, taking the absorbance extinction coefficient as prior knowledge. From the excess molar refractive index signal compared to the predicted refractive index signal based on the amino acid composition, along with the typical refractive index signal of carbohydrates (Part I, Appendix C), the average mass of carbohydrates per protein can be determined (Part I, Chapter 3, footnote on p. 94).

signal increment or the extinction coefficient at 280 nm may often be considered a safe prior knowledge for spectral analysis in $c_k(s)$.

The determination of signal coefficients is followed by the second, and uniquely useful application of the multi-signal model Eq. (7.1) in the study of mixtures. Here, the signal increments are usually predetermined and fixed. Signals need to be acquired at which the macromolecular components contribute differently. Detailed considerations of the choice of data acquisition wavelengths are discussed in Part I, Section 4.4. Briefly, besides the combination of the Rayleigh interference and absorbance optical system, data at different absorbance wavelengths may be acquired. Many proteins have significant absorbance in the visible wavelength range, such as many green, yellow, or red fluorescent proteins, as well as heme-containing proteins, but this is not the rule. However, proteins often differ significantly in their ratio of aromatic amino acid tryptophan and tyrosine, which determine their UV extinction spectra. As illustrated in Fig. 7.3, these amino acids differ considerably in their extinction ratio at the maximum at ~280 nm relative to the minimum at ~250 nm. SV data acquired simultaneously at both of these wavelengths is therefore often an excellent choice for the spectral deconvolution of different protein components.

Depending on loading concentration and signal/noise ratios, data at 230 nm can also usefully combined.

In some cases, naturally occurring spectral and refractive index differences are not sufficient to discriminate all components. In this case, the attachment of extrinsic labels may be necessary. This may be achieved, e.g., *via* chemical crosslinking of small molecule fluorophores to accessible cysteine or lysine residues,[4] site-specific enzymatic reactions, or expression of the protein in a fusion with a fluorescent protein. In either case, it is necessary to ascertain that the modified protein behaves identical to the native one with respect to the binding properties.

An important question is how different the component extinctions have to be. A quantitative measure introduced by Padrick and Brautigam [197] is termed D_{norm}. It measures how linearly independent the extinction vectors $\vec{\varepsilon_k} = (\varepsilon_{k,\lambda_1}, \varepsilon_{k,\lambda_2}, \ldots, \varepsilon_{k,\lambda_\Lambda})$ are among the different components, by determining he normalized volume of the parallelepiped formed by vectors $\vec{\varepsilon_k}$. If we arrange the extinction vectors into a $(K \times \Lambda)$-matrix $\mathbf{E} = (\vec{\varepsilon_1}, \ldots, \vec{\varepsilon_K})^T$, it is defined as

$$D_{\mathrm{norm}} = \frac{\det \mathbf{E}}{\prod_k |\vec{\varepsilon_k}|} \tag{7.2}$$

If the extinction vectors are completely orthogonal we have $D_{\mathrm{norm}} = 1$, and if they are linear dependent (proportional to each other), then $D_{\mathrm{norm}} = 0$. Usually, it takes values in between. Based on simulated SV data (with well-formed boundary properties and typical signal/noise ratio described in [197]) it was found that two components with D_{norm} less than 0.06 may not allow spectral discrimination.

Alternatively, we can carry out an analytical analysis of the propagation of data acquisition error *via* the spectral decomposition into component concentration error. This is largely equivalent but allows a more detailed consideration of the number of experimental data points resulting from species of given sedimentation coefficients and a given rotor speed, as well as experimental noise. As shown in Appendix B, in the spectral subspace for each s-value the $(K \times K)$ matrix $\mathbf{F} = \mathbf{E}\mathbf{E}^T$ governs the spectral resolution, in a linear equation system with elements

$$F_{k,\kappa} = \sum_\lambda \varepsilon_{k,\lambda} \varepsilon_{\kappa,\lambda} . \tag{7.3}$$

The error amplification factor may be estimated from the condition number of this matrix as

$$\frac{\sigma_c}{c} \approx \frac{\sigma_a}{a} \frac{\mathrm{cond}(\mathbf{F})}{\sqrt{n_{\mathrm{tot}}}} . \tag{7.4}$$

This allows us to predict, for particular sedimentation conditions, the error amplification from %-error in data acquisition to a %-error in component concentrations of species with the same s-value.

---

[4]When using this option, it is generally desirable to attach chromophores that are not too hydrophobic. For example, 5,6-caraboxyfluorescein (FAM) has been associated with artifacts modulating existing binding energies or creating new binding modes [200, 201].

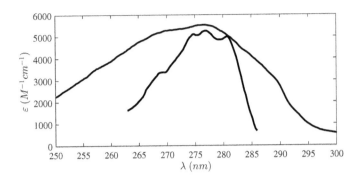

Figure 7.3 Comparison of the molar extinction coefficients of tryptophan (blue) and tyrosine (black, 4-fold amplified). Data taken from NIST Chemistry Webbook (http://webbook.nist.gov/chemistry/).

Both measures of spectral discrimination, D_{norm} and σ_c/c are implemented in SEDPHAT, and can be found in the Options ▷ Interaction Calculator function MSSV Condition and Dnorm Calculator. The condition calculator will determine the expected number of meaningful data points depending on the sedimentation condition and s-value of interest.

7.1.3 Examples: Enhanced Hydrodynamic Resolution from Spectral Decomposition

A first example of the potential of MSSV is shown in Fig. 7.4. A mixture of three proteins containing different fractions of aromatic amino acids was scanned simultaneously by UV absorbance at 280 nm and 250 nm, and additionally, the refractive index-sensitive Rayleigh interference data were acquired. In a standard $c(s)$ analysis the different signal contributions of each protein are reflected in the different peak heights and shapes, but it would be very hard to identify the sedimentation properties of each component. This is true, in particular, due to the very close sedimentation coefficient of two of the proteins, producing only two resolved main peaks in $c(s)$. However, extinction coefficients $\varepsilon_{k,\lambda}$ can be introduced, determined from a similar experiment carried out side-by-side on the individual components, as in Fig. 7.2. The global MSSV analysis now reveals the nature of the peaks. Notably, the broad fast peak in the $c(s)$ distribution now decomposes into a slightly faster and a slightly slower component $c_k(s)$. This demonstrates how spectral resolution in MSSV can improve the hydrodynamic resolution.[5]

---

[5] A simpler approach to exploit spectral differences would be to calculate sedimentation coefficient distributions for each signal first, such as in Fig. 7.4 Panels B,D, and F, and then to apply a secondary spectral analysis at each s-value, based on the $c(s)$-values at the different signals. This would be successful only in simple cases with perfect hydrodynamic resolution. Due to the need for regularization, which causes peak shapes and widths in the standard $c(s)$ analysis to vary dependent on the signal/noise level, artifacts in the spectral decomposition would arise. For example,

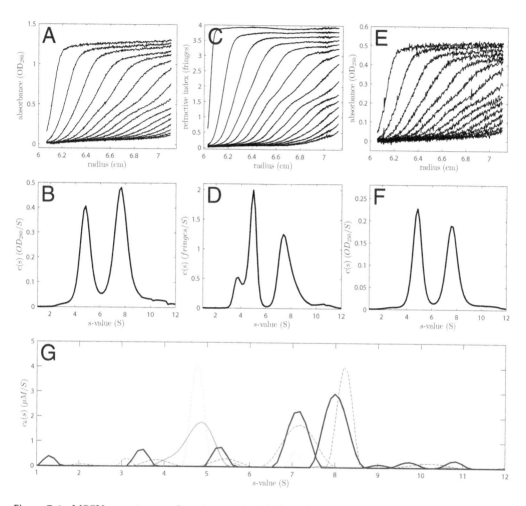

Figure 7.4 MSSV experiment of a mixture of BSA (green), IgG (red), and aldolase (blue) observed simultaneously by absorbance at 280 nm (*Panel A*), by Rayleigh interference (*Panel C*), and absorbance at 250 nm (*Panel E*) [35]. *Panels B, D*, and *F*: The individual $c(s)$ analysis of the different SV data sets at each signal displays only two major peaks ($c(s)$ traces correspond to the data in the panel above). Although different peak shapes and area ratios can be discerned, from the $c(s)$ data, the nature of the peaks is unclear. *Panel G*: Using pre-determined extinction coefficients of each component (calculated from separate MSSV experiments with the individual protein components such as in Fig. 7.2), the identity of the peaks can be unraveled in the joint MSSV analysis of the mixture. From the $c_k(s)$ distributions (solid lines, showing the BSA contribution in a 5-fold reduced scale) it can be discerned that the slower peak is that of BSA (green), while the faster peak is a combination of IgG (red) and the slightly faster sedimenting aldolase (blue). For comparison, the $c_k(s)$ from the individual protein samples measured separately are shown as dotted lines.

As mentioned above, the ability to determine the composition of hydrodynamically resolved complexes is one of the natural applications of MSSV. One example showing a 4:2 composition of a protein complex was already shown Part I, Chapter 4, Fig. 4.37. A more complex mixture of three interacting immune receptors can be found in Fig. 7.5.

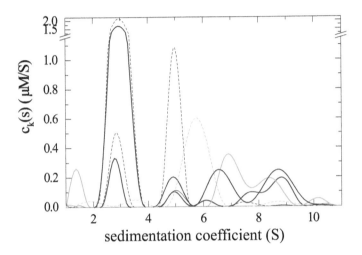

Figure 7.5 Example of the multisignal $c_k(s)$ analysis of a triple protein mixture of a viral glycoprotein (green), its cognate receptor (blue), and a heterogeneous antigen-recognition receptor fragment (red). The content of each protein component in the different s-ranges is obtained from the global analysis of sedimentation data acquired with the interference optics and with the absorbance system at two different wavelengths (data not shown), using two chromophorically labeled proteins and one unlabeled protein. Solid lines show the $c_k(s)$ analysis of the triple mixture (plotted in units of μM/S, with an axis break of the ordinate at 1.1 to highlight species in lower abundance). The analogous distributions of each protein alone are shown as dashed lines. The formation of two coexisting binary complexes at ∼5 S and ∼7 S and a ternary complex with 1:1:1 stoichiometry at ∼8.5 S can be discerned. Figure reproduced from [202].

7.1.4 Enhanced Spectral Decomposition from Hydrodynamic Considerations

Based on the principles outlined so far for the combined hydrodynamic and spectral analysis of SV data, it is straightforward to enhance the capabilities of MSSV further. The multi-signal $c_k(s)$ analysis can be applied to a segmented description of the range of s-values, a strategy that was described above for the hybrid discrete/continuous single-signal sedimentation coefficient distribution $c(s)$ in

---

non-constant $c(s)$ ratios would be found across a peak even for a single species, and simplistic interpretation as a linear combination of extinction coefficients would result in fluctuating values (and may imply negative concentrations). Restriction of a second-stage spectral analysis to $c(s)$ peak areas of each signal would be more advantageous, but still fail in cases of poor hydrodynamic resolution such as in Fig. 7.4. If second-stage spectral decomposition is carried out on distributions that do not provide an adequate boundary model, such as $g(s^*)$ from dc/dt (compare Fig. 6.5), problems are likely exacerbated [193].

Eq. (5.10). Extending the previous approach in the context of MSSV, each segment and discrete species can be assigned different spectral properties. Imagine the assembly of two different small proteins into large mixed complexes. With this in mind, we may combine the spectra of the components (or "chromophores") k into complex spectra κ of higher-order species. Specifically, we can express the signal increments $\varepsilon_{\kappa\lambda}$ for complexes as stoichiometric combinations of the signal increments $\varepsilon_{k\lambda}$ of the components, with the stoichiometry S_k^κ that reflects the number of subunits k in the complex κ[6]

$$\varepsilon_{\kappa\lambda} = \sum_k^K S_k^\kappa \varepsilon_{k\lambda} . \tag{7.5}$$

A total of Λ signals can determine a total of $K \leq \Lambda$ complexes, just like it could determine K components, if the stoichiometry matrix S_k^κ is full rank, i.e., the composition vectors are linearly independent. The complex signals $\varepsilon_{\kappa\lambda}$ can now replace the component spectra $\varepsilon_{k\lambda}$ in the MSSV decomposition. This is useful after we break the single continuous s range of Eq. (7.1) into different segments, in a multi-signal extension of the hybrid discrete/continuous model [35]:

$$a_\lambda(r, t) \cong b_\lambda(r) + \beta_\lambda(t) + \sum_{i=1}^{I} \sum_{\kappa_i=1}^{K} C_{\kappa_i}^i \varepsilon_{\kappa_i\lambda} \chi_1(s_i, D_i, r, t)$$

$$+ \sum_{j=1}^{J} \sum_{\kappa_j=1}^{K} \varepsilon_{\kappa_j\lambda} \int_{s_{\min}}^{s_{\max}} c(s)\chi_1(s, D^{(j)}(s), r, t)ds , \tag{7.6}$$

where, analogous to Eq. (5.10), I is the number of discrete species and J the number of continuous s-segments, each of which may describe up to K spectrally different species κ. Each of these, in turn, may be of different composition S_k^κ of the "basic component spectra" k following Eq. (7.5) or may have entirely different spectral properties — even those that are spectrally not distinct from the component spectra if they are hydrodynamically sufficiently distinct.

This seemingly complex model simply reflects the fact that, in light of hydrodynamic separation of species, we may pick different bases for the spectral decomposition in different s-ranges. Complementary to the enhanced hydrodynamic resolution due to spectral discrimination in Fig. 7.4, essentially now the hydrodynamic resolution feeds back to provide spectral constraints or spectral discrimination.[7] This is extremely useful in practice, since, for example, buffer signals offset the sedimentation data in the range below 1 S, which cannot be reasonably described in terms of protein component spectra, whereas there might be a range of s-values where only monomeric free protein species exist (with separately pre-determined

---

[6]It should be noted that this relationship also implies new molar units for the complexes.

[7]An example of this principle is the determination of absorbance spectra for hydrodynamically distinct species in [193].

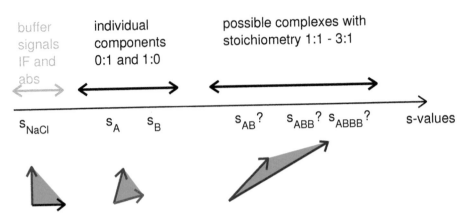

Figure 7.6 Schematics of segmented spectral decomposition adjusted to different s-ranges. Spectral decomposition is schematically drawn as colored vectors below each s-range, with a green patch indicating the space of spectral signals that is modeled. For higher complexes the stoichiometry constraints translate into corresponding spectral constraints which can be applied within the s-range of complexes. In the example shown, the assumption is embedded that complexes may have a single copy of A and between 1 and 3 copies of B. By contrast, it is not allowed in this highest s-range to find molecules with the spectral signature of free A or free B.

spectral properties), and another size range at higher s-values that is only occupied by mixed complexes of a certain stoichiometry.

Fig. 7.6 illustrates this aspect schematically for an SV analysis of data taken with two signals, e.g., interference and absorbance optics, introducing a vector approach to describe the spectral decomposition: If we draw the signal contribution to the interference signal in the vertical direction and contributions to the absorbance signal in the horizontal, we can assign each combination of signal increments ε_λ a vector in this plane. For example, signals from sodium chloride have only a refractive index contribution, and its vector would point vertically upward. On the other hand, signals from molecules that have very small refractive index increments but large extinction coefficients, such as nucleotides, and their vector will be close to horizontal. In general, the spectral components will have vectors $\varepsilon_{A,\lambda}$ and $\varepsilon_{B,\lambda}$ that point to some angle. In this picture, the requirement for the spectral contributions to be linearly independent simply states that the vectors of the components must not be parallel.[8] However, the angle between the spectral basis vectors can be different in different segments of the $c_k(s)$ distribution, with the non-negativity of concentrations embedding powerful constraints into the analysis by excluding spectral space. In the example of Fig. 7.6 the assumption is made that a single copy of A may bind between 1 and 3 copies of B — a prior knowledge that may derive, for example, from structural considerations. This excludes species in the s-range of the complexes (i.e., above that of the free species) with a spectral signature

---

[8]That is for two signals — for three signals they also must not be in the same plane.

of free **A** or free **B**, and clearly any species that contributes solely to interference or solely to absorbance signals, respectively.

The segmentation of the $c_k(s)$ distribution is motivated by prior knowledge from independent data and from structural considerations. It is also possible to embed a hypothesis into the analysis by allowing only species of a certain spectral signature (for example, species with 1:1 molar ratio) to occur in a certain segment.[9] Comparison of the quality of fit obtained with different spectral constraints can provide a criterion to decide between different interaction models. Segmenting the range of s-values can also allow us to fit for the extinction coefficients of the complexes (if sedimenting in a separate boundary), thereby accounting for possible hyper- or hypochromicity effects. Another powerful strategy resting on Eq. (7.6) is the representation of free species in the model as discrete species, where possible. This allows reducing the potential correlation between the different regions of s-values.

SEDPHAT offers the same interface for segmented $c_k(s)$ distributions with regard to the partitioning of s-values as for the standard hybrid discrete/continuous distributions at a single signal. For each segment, multiple spectrally different $c_k(s)$ branches can be specified. Spectra of continuous segments and discrete species can be entered directly as signal coefficients $\varepsilon_{\kappa,\lambda}$, or *via* their stoichiometry matrix S_k^κ in multiples of basic "chromophore" spectra $\varepsilon_{k,\lambda}$. The latter can be refined in non-linear regression.

7.1.5 Spectral Regularization by Mass Conservation

In some cases different components may not be sufficiently spectrally distinct to achieve reliable MSSV decomposition, and signals may be mis-assigned. This leads to a situation where significantly more of one component seems to be reflected in the $c_k(s)$ distribution than what was loaded. In this case it can be very helpful to introduce independent information on the total loading concentration of one or more components into the data analysis [37]. To this end, experiments should be designed with suitable control experiments of individual components to allow the determination of the concentration of stock solutions. With careful pipetting — absent significant co-precipitation[10] — mixtures can then be analyzed with constraints in the total amount of material of each component. This known total concentration can be introduced in the analysis as a condition for the integral over the component $c_k(s)$ in a certain s range to match the known total loading concentrations C_k^{tot} [37]:

---

[9]It should be kept in mind, however, that reaction boundaries may have non-integral molar ratios [75].

[10]Conceivably co-precipitation may be flagged by a loss of concentration in both components, as opposed to spectral mis-assignment that results in an overestimate of one and underestimate of another component.

$$C_k^{\text{tot}} \cong \int_{s_{k,\text{MC},1}}^{s_{k,\text{MC},2}} c_k(s) ds \tag{7.7a}$$

$$a_\lambda(r,t) \cong \sum_{k=1}^{K} d\varepsilon_{k,\lambda} \int_{s_{\min}}^{s_{\max}} c_k(s)\chi_1(s, D_k(s), r, t) ds + b_\lambda(r) + \beta_\lambda(t). \tag{7.7b}$$

Here, both equations are simultaneously matched by least squares. This is also re-ferred to as MC-MSSV [37]. $s_{k,\text{MC},1}$ and $s_{k,\text{MC},2}$ are the minimum and maximum s-values, respectively, where any of the components are expected to sediment, for the purpose of assessing mass conservation. Mathematically, as outlined in Appendix B, mass conservation can improve the conditioning of the spectral decomposition by replacing the relevant extinction coefficient matrix $(\mathbf{F})_{k,\kappa}$ from Eq. (7.3) by $(\mathbf{F} + \alpha'\mathbf{I})_{k,\kappa}$, where $\mathbf{I}$ is the identity matrix and α' is a scaling factor. Thus, mass conservation information can effectively substitute for insufficient spectral discrim-ination of components.

Clearly the mass conservation condition (7.7a) will generally conflict with find-ing the best least-squares fit to the raw SV data (7.7b), and a balance has to be found. Accordingly, mass conservation can be imposed in two opposite ways: (1) the mass conservation condition is used solely as a regularization term, much like Tikhonov or maximum entropy regularization in the s dimension, and deviations are accepted to avoid creating a statistically significant decrease in the quality of fit; (2) mass conservation is strictly enforced to within a certain tolerance (e.g., allowing for a possible pipetting error of 5%) even if it causes a significant misfit to the SV data. Ideally, both approaches will be consistent.

Mass conservation constraints for MC-MSSV in SEDPHAT can be entered in the lower right section of the MSSV global parameter box. In order to provide flexibility for bal-ancing mass conservation constraints in multiple components, the checkbox penalize % defect allows us to assess deviations either in absolute concentration (unchecked), or in % of the total concentration (checked). Mass conservation can be used purely as regularization, not allowing a significant decrease of fit quality (auto adjust by chisq, P), or be enforced to within a certain tolerance irrespective of a worsening fit (auto adjust enforce to within (%)).

A limitation of MC-MSSV is that the mass conservation condition cannot gen-erate information about the s-value the component is sedimenting within the con-sidered integration range $[s_{\text{MC},1}, s_{\text{MC},2}]$. Therefore, even though MC-MSSV can be carried out even without any spectral discrimination, it will be possible to derive only the average molar ratios of components across the range $[s_{\text{MC},1}, s_{\text{MC},2}]$. This limitation can be alleviated if the s-range can be partitioned such that one com-ponent is completely excluded in a certain range. For example, if from separate experiments it is known that $s_A < s_B$, then B can be excluded from the s-range of A. Similarly, theoretical considerations in the effective particle model for interacting systems can limit the possible s-range of certain components [37].[11]

7.2 MONO-CHROMATIC MULTI-COMPONENT DISTRIBUTIONS

7.2.1 Temporal Signal Modulation

Quite analogous to the multi-component sedimentation coefficient distributions $c_k(s)$ in MSSV, it is possible to unravel different co-sedimenting components that are not spectrally distinct but offer characteristically temporally modulated signals [41]. As introduced in Section 1.1.3.1, reversibly photoswitchable fluorophores can produce, in suitable illumination conditions in the analytical ultracentrifuge, strongly time-dependent fluorescence and absorbance signal magnitudes, which may involve both signal decrease or increase. The modulation may occur due to the incident photons from the scanning laser in the fluorescence detection system leading to gradual switching as in Eq. (1.4), or from pulsed illumination of an additional light source creating a virtually instantaneous jump [41].

We may capture temporally changing signals as time-dependent signal coefficients $\varepsilon_k(t)$. However, in a more detailed consideration described above in Eq. (1.5) and more extensively in Part I, Section 4.3.2.3, we find that the photon flux incident on the sedimenting molecules when using the fluorescence detection system will depend on radius position, and that the total exposure depends on the spatial trajectory and thereby its s-value [41]. Therefore, a more complete description of the signal coefficients for photoswitchable molecules is $\varepsilon_k(s, r, t)$. Analogous to the spectral multi-signal case, the specific temporal signature should be measured in separate experiments with the individual components.

The temporally modulated mono-chromatic multi-component (MCMC) distribution is based on the description of the data as a superposition of sedimentation patterns from K_T components with such temporally distinguishable signals

$$a(r,t) \cong \sum_{k=1}^{K_T} \int_{s_{min}}^{s_{max}} c_k(s)\varepsilon_k(r,s,t)\chi_1(s, D_k(s), r, t)ds. \tag{7.8}$$

As outlined in Appendix B, in near complete analogy to the multi-signal analysis, the resolution of different components is governed by a matrix similar to $\mathbf{F}$ in Eq. (7.3) but now composed of sums over time points of scans rather than sums over spectral signals [41]

$$\begin{aligned} \mathbf{F}_{k,\kappa} &= \sum_t \varepsilon_k(s,r,t)\varepsilon_\kappa(s,r,t) \sum_r \chi_1(s,r,t)^2 \\ &\approx \sum_t \varepsilon'_k(t)\varepsilon'_\kappa(t)w(s,t), \end{aligned} \tag{7.9}$$

---

[11]To this end, experimental configurations are beneficial where the smallest component is in a molar excess, which ensures they are the constituent of the undisturbed boundary [37].

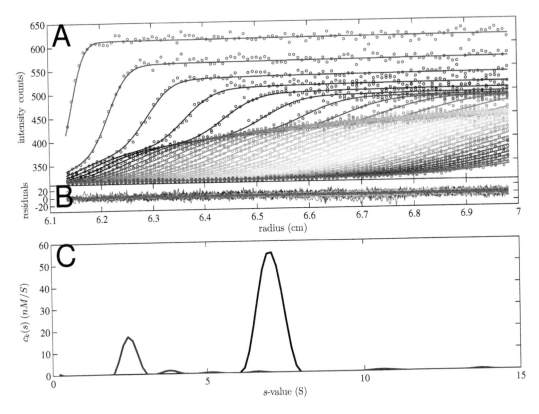

Figure 7.7 MCMC analysis of fluorescence optical SV data acquired with excitation at 488 nm of a mixture of 20 nM rsEGFP2, 20 nM Padron, and 50 nM anti-EGFP mAb. Data reproduced from [41]. *Panel A*: Radial fluorescence scans (circles, showing only 3rd data point in every 2nd scan). The MCMC fit (solid lines) accounts for two components, one with increasing signal for Padron, and one with decreasing signal for rsEGFP2. *Panel B*: Residuals of the fit. *Panel C*: Calculated component sedimentation coefficient distributions for padron (magenta) and rsEGFP2 (blue).

where the second equation takes advantage of the only weak s- and r-dependence of the signal coefficient, $\varepsilon_k(s,r,t) \approx \varepsilon_k'(t)$, and the abbreviation $w = \sum_r \chi_1(s,r,t)^2$.

The latter expresses the fact that temporal signals can only contribute to component discrimination if they are reflected in a significant number of data points. For example, if a signal decay is very rapid relative to sedimentation, then the number of meaningful data points reporting on the sedimenting species is small (unless the signal is periodically reset during the sedimentation process). If it is too slow, then rapidly sedimenting species may not produce a significant signal change before the boundaries are outside the radial observation window [41].

Inspection of Eq. (7.9) shows that perfect discrimination in MCMC would be achieved when the cross products between different components vanish, for example, when one signal is always off while the other is on. This is formally completely analogous to the ideal case of spectrally exclusive detection in MSSV where data

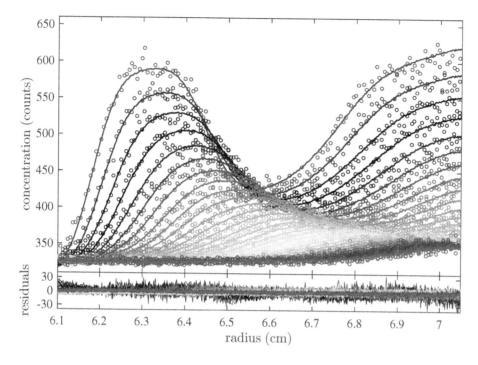

Figure 7.8 SV data of a photoswitchable molecule that has been locally converted into a dark state by stationary exposure to the excitation beam for 10 minutes at the beginning of the run. This creates a spatio-temporal pattern that can be easily distinguished from constant signal contributions of non-photo-convertible molecules in a mixture by MCMC analysis. Data reproduced from [41].

acquisition alternates between different wavelengths. In fact, there is no reason temporally modulated signals could not be acquired at different signals and combined into a multi-dimensional global analysis that should allow $K \times K_T$ components to be distinguished. Similarly, the framework is fully compatible with mass conservation regularization described for MSSV in Section 7.1.5.

7.2.2 Spatio-Temporal Signal Modulation

With photo-switchable molecules it is possible to extend the SV analysis even further with spatially selective photoswitching [41]. Akin to FRAP, for example, it is possible to induce strong photoconversion over a narrow region in the sample solution for a limited period of time. The remaining signal exhibits a trough across this region, which will migrate and be leveled by diffusion with time. In fact, even during the localized "bleaching" process, sedimentation and diffusion are taking place and contributing to the shape of the trough. An example is shown in Fig. 7.8; it is easily discerned how the early localized photoconversion creates two additional signal boundaries — despite the absence of a concentration boundary — which significantly contribute to the quantitation and characterization of the photoswitchable component [41]. This is best theoretically described and computationally implemented as a coupled sedimentation/diffusion/photoswitching process between species in a visible state, $\chi_v(r,t)$, and a dark state, $\chi_d(r,t)$:

$$\frac{\partial \chi_v}{\partial t} = -\frac{1}{r}\frac{\partial}{\partial r}\left(\chi_v s \omega^2 r^2 - D\frac{\partial \chi_v}{\partial r}r\right) - \chi_v \phi_{v\to d}(r,t) + \chi_d \phi_{d\to v}(r,t)$$

$$\frac{\partial \chi_d}{\partial t} = -\frac{1}{r}\frac{\partial}{\partial r}\left(\chi_d s \omega^2 r^2 - D\frac{\partial \chi_d}{\partial r}r\right) + \chi_v \phi_{v\to d}(r,t) - \chi_d \phi_{d\to v}(r,t)$$

(7.10)

with $\phi_{v\to d}(r,t)$ and $\phi_{d\to v}(r,t)$ the radial- and time-dependent probabilities for conversion from the visible to dark state and reverse, respectively, as controlled by illumination conditions. Obviously, the visible state is the only one contributing to the signal, which in fluorescence detection will undergo the same scanning-induced time dependence as described in Eq. (7.8), and constitutes one component in the MCMC analysis.

MCMC can be carried out with the SEDPHAT model `mono-chromatic multi-component fluorescence distribution analysis`. In complete analogy to MSSV, MCMC is implemented in SEDPHAT with the same possibilities for segmentation of the s-value range as the hybrid discrete/continuous distributions. The temporal signal dimension can be modeled with exponentials in combination with up to three "reset" events that each will restore the signal to a certain % level of the initial signal. Currently up to three different components can be specified. The resolution for the given temporal signatures (not including spatial patterning) can be predicted according to Eq. (7.9) with the function `Options ▷ Interaction Calculator ▷ MCMC Condition Calculator`, predicting the meaningful number of data points w for a given s-value and SV run conditions.

CHAPTER 8

Practical Analysis of Non-Interacting Systems

T HE present chapter discusses several strategies and considerations for the practical application of the analysis principles outlined above. Throughout, we assume that a successful sedimentation experiment was carried out, as described in Part I [1], showing convection-free sedimentation at an adequate rotor speed, observed over a sufficient period of time with a suitable selection of optical detection. Further, we assume that the non-interacting species and distribution model is really applicable, either because the system under study is practically non-interacting (which can usually be verified through experiments in a concentration series), or because the non-interacting system model is used as an approximation, for example, for an effective sedimenting particle of an interacting system [70, 75, 76], or as a vehicle for the determination of a signal weighted-average sedimentation coefficient. Sedimentation processes of systems with attractive or repulsive interactions will be the topic of a forthcoming volume, but many of the principles discussed here will also apply. The following will focus on strategies that address the key problem of SV of non-interacting systems, which is to distinguish the extent of diffusional boundary spread from the differential migration arising from different size particles.

8.1 GENERAL FITTING STRATEGIES

8.1.1 Data Preprocessing

Ordinarily there should be no data preprocessing required, because all analysis parameters relating to sedimentation and optical signal are explicitly part of the fitting model. However, before a fitting session can start, the following operations may have to be carried out:

data conversion The standard mode of data acquisition when using the absorbance scanner (Part I, Section 4.3) is in absorbance mode. If data are collected in intensity mode the scan files need to be converted to pseudo-absorbance data and saved in the standard absorbance format separately for each sector.

file sorting Some situations require the identification of suitable file types for analysis. For example, when using the absorbance scanner at multiple wavelengths the filename is incremented with each scan and does not reveal the data acquisition wavelength. A SEDFIT function Options ▷ Loading Options and Tools ▷ Save Raw Data Set allows us to change the filenames so that the wavelength becomes part of the filename. In this way, sets of equivalent scans can be picked for loading in SEDFIT or SEDPHAT.

Similarly, when working with fluorescence optical data, a very large number of scans may be generated from a single experiment, due to the availability of up to 14 sample sectors, each of which may be scanned at multiple gain and photomultiplier voltage settings. Often, only data from certain settings within a certain time range are of interest. A SEDFIT utility function in the Options ▷ Fluorescence Tools menu can browse through existing data files, apply user-defined filters for scan selection, and copy the condensed data into a new folder along with pre-made list files.

scan time Due to errors in the recorded scan times in the file headers generated by the ultracentrifuge operating software [50, 51] it is recommended to read time intervals between scans from the computer operating system file timestamps and create corrected scan files. This issue has been discussed in detail in Part I, Section 4.2.4.

speedsteps When analyzing scans acquired at different rotor speeds, for example from the use of a time-varying rotor speed schedule, the time course of rotor speed needs to be established. This can be done manually by creating an ASCII file speedsteps.txt using the format outlined in Section 2.2.4, or by extracting this information from preliminarily loaded scan files in SEDFIT (ensuring at least one scan from each rotor speed step is present).

unstretch For analyses with time-varying centrifugal field models, it is also important to compensate for differences in the radial position of the solution column caused by rotor stretch using the known stretching modulus implemented in SEDFIT (see Part I, Section 5.2.3.2). Scan files need to be saved with translated radial scale and corrected rotor speed to compensate for changes in the radial-dependent centrifugal field.

Generally not all of these operations are necessary. If more than one is used, they should be executed in the order listed above.[1]

---

[1]Rare experimental, instrument, or computer glitches may require additional pre-processing steps, a number of which can be found in the Options ▷ Loading Options and Tools menu of SEDFIT.

8.1.2   Time Range of Selected Scans

Is important that the selection of scans included in the data analysis evenly reflects the entire sedimentation process. Scans from time points long after sedimentation is completed should be avoided, unless they report on small molecules still slowly migrating. This is because too many repetitive scans of the final baseline would lead to an unnecessary increase in computation time, diminish the relative depth of minima in the error surface of best-fit parameters, and artificially appear to improve the quality of the fit if assessed solely by the rmsd or residuals histogram. Likewise, it is dangerous and can cause misleading results to exclude the earliest scans, even though they might be the most difficult ones to fit.

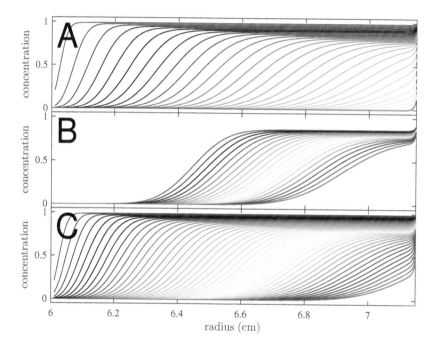

Figure 8.1   Examples for the selection of scans for data analysis. *Panel A*: Inclusion of too many late scans that reflect only the baseline. This is poor practice (unless working at very low signal/noise ratio) due to the redundancy of scans that reflect only the baseline. *Panel B*: A subset of scans within a narrow time window: Lack of early and late scans truncates information on diffusion and polydispersity. *Panel C*: An even representation of the sedimentation process is usually best.

There is a straightforward method to arrive at a good choice of scans for the analysis after an initial survey in **SEDFIT**. First, we select all acquired scans (of one particular cell) to be loaded. If there are more than 100 scans available, we can use the option to only load every n^{th} scan to reduce the total number of loaded scans to between 30 and 100.[2] This may look like Panel A of Fig. 8.1. The color coding of scans in **SEDFIT** uses increasing color temperature mapped to the sequence of the

---

[2]For lack of a compelling reason to load more, there is an upper limit of 1000 scans that can be loaded in **SEDFIT**.

loaded scans in order of their elapsed time in the sedimentation process. Therefore, if only blue scans are visible, and yellow, orange, and red are all compressed, this signifies that too many late scans have been added. Hovering over boundaries with the mouse will display the scan number, so that an appropriate upper limit can be estimated, such as reflected in the data set of Fig. 8.1C. The choice depicted in Fig. 8.1B is suboptimal since it severely truncates the opportunity to discriminate polydispersity and diffusion.

An exception to this principle of evenly representing the entire sedimentation process occurs for data at a very low signal/noise ratio. In this case, it can be very helpful to additionally include many late scans to overrepresent the baseline, such that boundaries with signal magnitudes on the order of the noise in the data acquisition can be recognized with good statistical confidence [203,204]. Conversely, if a species sediments much more rapidly than all others, it may be necessary to include all available early scans.

> The most common loading function Data▷Load New Files allows to either select scan files followed by the option to load every n^{th} scan, or a list file stemming from a previously loaded data set. The function Data▷Add More Files is helpful, for example, to increase the number of early scans. Another option to refine the data file selection is manually editing the list file in a text editor.

The number of scans loaded has direct bearing on the computation time. The computation time to fit sedimentation boundaries with distribution models is proportional to the total number of data points (Appendix B). It is possible — though not commonly necessary — to reduce the number of data points per scan by averaging neighboring data points in present radial intervals. This may be an effective strategy, at least for a first analysis, to speed up the analysis when working with interference optical data. These data have a very high radial density of points ($\Delta r = 0.0007$ cm), and in the absence of steep boundaries, the radial density can be lowered without significant loss of information. In contrast to loading only every n^{th} scan, where scans are simply left out, here the information from each data point is honored in the pre-averaging process leading to lower statistical noise parameters. An alternative and preferred approach to keep computational time low is the use of multi-threading, which divides up the main costly steps into different processes executed in parallel.

The radial density of data points when loading scans in SEDFIT is controlled by the function Options ▷ Loading Options and Tools ▷ set radial resolution. The default value of 0.001 cm is interpreted as an instruction to load all data (even if the radial resolution exceeds 0.001 cm). Values higher than 0.001 cm will cause radial pre-averaging of data points while loading, to reduce the total number of points.

Multi-threading is controlled in the function Options ▷ Size Distribution Options ▷ compute with multiple threads. In SEDPHAT, the same function is located in Options ▷ Compute With Multiple Threads. This is a toggle, and when switched on, the number of threads can be set, corresponding to the total number of processing cores available in the computer used. If this is saved as a default setting, then it does not have to be specified again. However, when instances of SEDFIT or SEDPHAT are started from executable files located within a different path, these will not retrieve the same configuration parameters, and therefore multi-threading levels need to be established separately.

8.1.3 Meniscus and Bottom

The meniscus and bottom position are important landmarks to be identified after scans have been loaded. The appearance of the meniscus and bottom of SV experiments in different optical systems has already been described in great detail in Part I, Section 4.1.2, and it is important to be familiar with their appearance. The meniscus of the solution column generates an optical artifact over a relatively wide radial range, with a shape that is dependent on a number of factors, including rotor speed, surface tension and solution composition, solution refractive index, and optical focus [205]. Most importantly, it must be recognized as an optical artifact that no standard AUC optical detection system is designed to image. Historically, SV analysis methods have often required a meniscus position determined *ad hoc* from the graphical inspection of the location and shape of the optical artifacts (including the van Holde–Weischet and time-derivative methods). This appeared to work well, perhaps partly due to the lack of quality control *via* inspection of the residuals between the implied analysis model and the experimental data. However, for a more detailed boundary modeling, this is unsatisfactory (Part I, Section 4.1.2.4).

Fortunately, the value of the meniscus radius is not necessary to be fixed for the interpretation of the SV data by direct boundary modeling. Rather, sedimentation analysis can rely solely on the progression of the boundary positions with time, and the meniscus arises merely as an implicit projection back to zero time [10, 11, 57]. The best-fit meniscus position is usually very well determined, with a precision that can exceed the radial resolution of the data points [52]. An exception is the analysis of data with very low signal/noise ratio, which do not have sufficient information content for the meniscus position to be defined, such that a graphical estimate must be used.

Despite the computational refinement, the graphical estimate of the meniscus position is still of interest, but, aside from serving as an initial guess for the fitting parameter, it is predominantly informational. It can aid in the quality control of

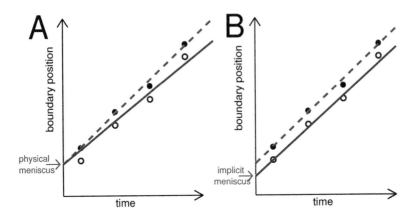

Figure 8.2 Schematic representation of boundary positions as a function of time in an ideal experiment (solid black circles) and in the case of initial temperature-driven convection or other experimental imperfections leading to an initial delay followed by normal migration (open circles). In this schematic, the slope of the best-fit line represents the estimated sedimentation coefficient, which would apply in the rectangular approximation of sedimentation. *Panel A*: The physical meniscus is estimated (e.g., from an overlay graph of all scans) and used as a fixed point in the data analysis. This constrains the fitted line in the origin at $t = 0$. At best — disregarding optical limitations — this will work in an ideal experiment (red dashed line). However, in case of initial convection, the constraint will impose an erroneous slope in the best-fit line (magenta line). *Panel B*: Here, no assumption is made on the origin of the boundary at $t = 0$. As a result, the red and magenta lines are parallel, and the slope of the line through the imperfect data (open symbols and magenta line) is close to the ideal case (red dashed line). The implicit meniscus position differs from the physical meniscus position, but the sedimentation coefficient is a better estimate [38].

the fit to flag the presence of significant experimental imperfections. Similarly, an inadequate boundary model may cause significant deviations between the implicit best-fit and graphically estimated meniscus location. To this end, it is useful to graphically define the uncertainty of the physical meniscus position from the edges of the artifact-free data contained fully in the air space above the solution column and fully inside the solution column, respectively (see Fig. 4.3 in Part I). One may use these initially as a constraint in the computational refinement of the meniscus. However, it can be advantageous for obtaining more precise s-values if the meniscus position is allowed to assume the best position implied by the boundary progression, even if there is a discrepancy with the appearance of the meniscus in the data scans: In this way, transient experimental imperfections, such as initial temperature-driven convection, will propagate less into large errors in s-values. This is illustrated in the schematics in Fig. 8.2, and was experimentally verified in [38].[3] The discrepancy between physical and implicit meniscus then signifies that the experiment should be repeated, if possible, to achieve greater accuracy.

---

[3]As shown in [38] and discussed in detail in Part I, this is preferable over the exclusion of early scans in the presence of initial convection — early scan exclusion will reduce the misfit but not reduce the errors in an analysis with graphically fixed meniscus.

A further advantage of treating the meniscus position as an adjustable parameter in the data analysis is that this allows the meniscus parameter to capture imperfections in the reported timing of the scans, which may suffer small constant time offsets for various reasons (Part I, Section 4.2). Among the many factors conceivably impacting on the precise timing of boundary migration are time required for the stabilization of the air/meniscus interface [206], effects of transient initial temperature and corresponding viscosity fluctuations from adiabatic cooling of the accelerating rotor [95], and compression of the solution column [207], all of which would, in first-order approximation, be accommodated as a shift of the effective best-fit meniscus position.

The meniscus is indicated in SEDFIT and SEDPHAT as a red vertical line through the data plot. It can be set either graphically by control-left-double-clicking the mouse at a position in the graph where the meniscus should be, or by dragging an existing red meniscus indicator line to a new position. All mouse actions need confirmation in a button that appears in the plot. Alternatively, the meniscus radius can be entered manually in the model parameter boxes in SEDFIT or the experimental parameter boxes in SEDPHAT. A checkbox next to the parameter indicates whether or not it is destined for refinement in non-linear regression.

Minimum and maximum radial values allowed in the refinement are also graphically represented and show as dottted vertical lines, which can be dragged to new positions with the mouse. Alternatively, meniscus constraints can be specified in the SEDFIT menu Options ▷ Fitting Options ▷ set fitting limits for meniscus, bottom, or in SEDPHAT after toggling the refinement flag (unmark the checkbox next to the meniscus, close the parameter box, reopen the parameter box, mark the checkbox for meniscus refinement, and enter values in the prompts following closing of the parameter box). For sharing meniscus parameters in SEDPHAT, flag the meniscus for refinement and enter the experiment number to which the meniscus is linked; it will then mirror these values (whether these are destined to be refined or not).

Identical operations control the bottom radius of the solution column, indicated by a blue line in the data plot.

An additional consideration arises in global analyses of data from the same solution column with multiple optical signals. Even though the imaged solution column, the boundary movement, and the elapsed time is truly identical, the calibration of radial positions will generally not exactly coincide (Part I, Chapter 6).[4] Therefore, shared numerical values for the meniscus location may map differently onto the radial scale of the solution column observed in different optical systems. As a consequence, shared (and jointly refined) meniscus positions are a highly useful constraint between data sets from the same cell acquired with the same optical system

---

[4]External radial calibration methods [51] are designed to achieve accurate radial magnification, but not absolute radial positions. The latter impact the accuracy of the derived sedimentation parameters much less, due to the long distance from the center of rotation (Part I, Section 4.1.1), but discrepancies in the absolute radial position calibration would directly impact meniscus constraints.

(e.g., different wavelengths in the absorbance scanner), but should be avoided across different optical systems.

The bottom position is usually not nearly as important as the meniscus — in fact, whether it plays any role at all depends on the extent of back-diffusion inside the radial fitting limits.[5] Without back-diffusion data to be fitted, the bottom position is irrelevant. This is fortunate since the bottom is even less well defined than the meniscus, for several reasons discussed in detail in Part I, Section 4.1.2.

If the model does depend on the bottom position, then the bottom radius must be treated as a parameter to be refined during non-linear regression. A constraint for the lower limit can be derived from the visual overlay of scans as the highest radius where accumulation of material can be imaged. The upper limit, however, is best set to a very high value, because surface adsorption processes may possibly diminish the extent of back-diffusion as if the bottom position was farther away from the data to be fitted than its physically realistic range, given cell and rotor geometry and factoring in possible calibration errors. An opposite effect, potentially making the bottom to appear at lower radii than physically possible, is conceivable considering repulsive non-ideality at the high concentrations near the interface that may exacerbate back-diffusion. These issues have not been studied in detail, and are likely system dependent. In any case, it is advisable — especially for material with steep gradients — to impose as few assumptions as possible on the sedimentation/diffusion process at the bottom interface, and therefore to adopt a wide range for the fitting limits of the bottom position.[6]

By default, for each species for which Lamm equation solutions are calculated as part of the model, SEDFIT will predict whether back-diffusion can possibly extend into the fitting range. If not, then a special set of boundary conditions for semi-infinite solution columns is applied (Appendix A) [34, 61]. This behavior is controlled in the Lamm Equation Options parameter box.

8.1.4   Radial Fitting Range

The data to be fitted usually extend from a minimum radius that is common to all scans up to a maximum radius.[7] The minimum radius for data to be fitted (the "left"

---

[5]However, in flotation processes as in [208], the roles of meniscus and bottom reverse.

[6]This does not apply for sedimentation equilibrium analyses, which rely entirely on shallow back-diffusion at low rotor speeds extending through the entire solution column. In sedimentation equilibrium mass conservation considerations relating equilibrium concentration profiles acquired at different rotor speeds allow a better definition of the bottom position [158] and the fitting range should be placed more judiciously.

[7]There is also the option to exclude a radial range of data in the middle of the scans, using the SEDFIT function Options ▷ Loading Options and Tools ▷ Exclude Region from Raw Data. However, there is rarely a need for this, since most imperfections can be very well accounted for by TI noise. For this reason, most numerical algorithms assume only a single lower and upper limit, and in the case of excluded data, require interpolation across the data gap.

fitting limit) should be as close to the meniscus as possible, but excluding optical artifacts in the vicinity of the meniscus. For absorbance optical data this is usually straightforward and typically leaves the left fitting limit at a radius ∼0.01 cm larger than the meniscus. In the case of interference optical data, an initial preliminary fit may be required to permit subtraction of an estimate of RI noise, such that the meniscus region can be visualized more clearly.[8] Often low-level oscillations occur within 0.01 cm of the interface, which should be excluded. For fluorescence optical data, depending on the focal depth, the optical aberrations may extend further into the solution column, and the width of the artifacts in the bottom region can be used as a guide for the distance between meniscus and left fitting limit (see Part I, Section 4.1.2.3).

Often, misfits arising from convection or incorrect models are largest close to the meniscus, but these residuals should not be used as a criterion for truncating the data to be modeled. On the other hand, optical artifacts within the radial analysis window will also cause residuals and can sometimes be identified in this way.

The limit of the highest radius of data to be included in the fit (the "right" fitting limit) is not only determined by the exclusion of optical artifacts, but also by a judgment on how much back-diffusion to exclude from the fit. This depends on several factors, including the molar mass of the smallest significantly populated species, the rotor speed, and the time range of the data set. Some examples are shown in Fig. 8.3. Panels A and B show cases where back-diffusion is absent or steep and localized close to the bottom, respectively, such that it can easily be excluded. If back-diffusion extends farther into the solution column, such as in Panel C, then omission of this region would sacrifice a significant portion of the data. Furthermore, the onset of back-diffusion from smaller species is not very steep, such that modeling is typically successful provided the bottom position is refined in the fit. In fact, similar to the Archibald method (Fig. 2.8 on p. 37), the inclusion of shallow back-diffusion will introduce additional information on the buoyant molar mass of the sedimenting species at all times. For example, in the $c(s)$ method this can aid significantly in the determination of the frictional ratios. This can be particularly useful in the study of broad boundaries, where otherwise diffusion and polydispersity are not well resolved. Finally, in the presence of very small species (Panel D), their back-diffusion extends across the entire solution column very quickly, and is impossible to exclude from consideration. Accordingly, along with the addition of parameters for the small species, the bottom position must be refined in the fit. However, the fitting limit can be set such that larger species' back-diffusion is still excluded.

---

[8]This can be achieved in the following way: (1) carry out a preliminary fit, (2) subtract both TI and RI noise estimates, (3) add back the TI noise estimate. This will leave the data approximately corrected for just the RI noise and otherwise show the original scans. Additional removal of TI noise is usually not advantageous, especially for inspection of the back-diffusion region, since the quality of the preliminary fit is typically not very good and the scans would be distorted by skewed TI noise profiles.

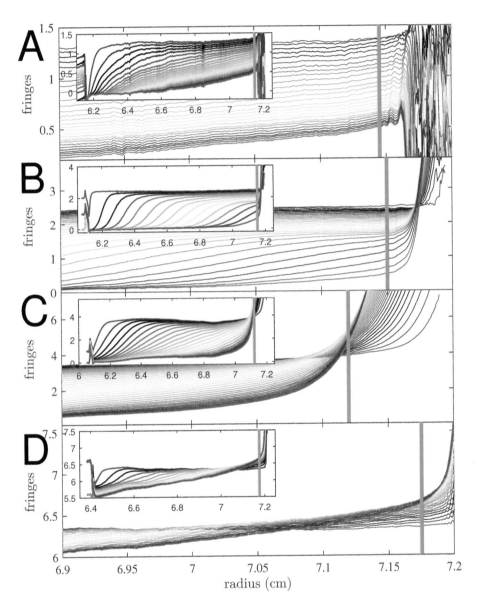

Figure 8.3 Examples of "right" fitting limits and back-diffusion. Panels A to D show radial fringe profiles close to the bottom, with the inset showing the entire data set. *Panel A*: Data set without back-diffusion. In this case, the fitting limit can be set to the highest radius excluding artifacts (such as oscillations) close to the bottom. *Panel B*: Back-diffusion is localized close to the bottom and shows very steep profiles. These should be excluded from the fit. *Panel C*: Back-diffusion extends relatively far into the cell. The complete exclusion of this zone would require sacrificing a significant portion of the data. It should be possible to model some the back-diffusion, since it is not very steep. This will improve information on the buoyant molar mass. *Panel D*: Here a sedimentation boundary of a larger species is superimposed by a sedimentation pattern of small molecules, the latter creating a shallow back-diffusion zone that extends quickly across the entire solution column. This must be quantitatively accounted for in the model. The fitting limit achieves exclusion of the back-diffusion of the larger species only.

8.1.4.1 Partial Boundary Modeling

If only a subset of the sedimentation coefficient range covered by the experimental data is of interest, it is possible to restrict the experimental data set subject to analysis with different radial fitting limits for each scan. In this technique, termed partial boundary modeling [38], the fitting limits are the theoretical radial positions, calculated according to Eq. (2.3), of non-diffusing species with sedimentation coefficients $s_{PB,min}$ and $s_{PB,max}$. In this way the fitting limits are equidistant in s-space relative to the boundary of interest. This allows the exclusion of signals from faster sedimenting aggregates, or clearly separated boundaries from minority species. Fig. 8.4 illustrates this approach showing a partial boundary model of the data of Fig. 5.8B on p. 107 that excludes the faster sedimenting species. However, it is important to recognize that exclusion of potential small species can usually not be achieved with partial boundary modeling due to the diffusion broadening of their boundaries [38].

When the limits $s_{PB,min}$ and $s_{PB,max}$ are set closely to focus exclusively on a single boundary in order to prepare data apparently suitable for a single-species Lamm equation model, there is an interesting analogy to the modeling of peaks in $g^*(s^*)$ with Gaussians for diffusing particles (Section 4.4): If one takes for $s_{PB,min}$ and $s_{PB,max}$, the same limits that one would choose for the fit of $g^*(s^*)$, then the partial boundary approach promises to exploit the same source of information from the experimental data, but without the drawbacks of the approximations inherent in both $g^*(s^*)$ and the Gaussian model Eq. (4.13) [38]. However, if the sedimentation boundary of interest harbors heterogeneity, then the attempt to use close range of $s_{PB,min}$ and $s_{PB,max}$ to focus on a single species would, of course, be similarly flawed.

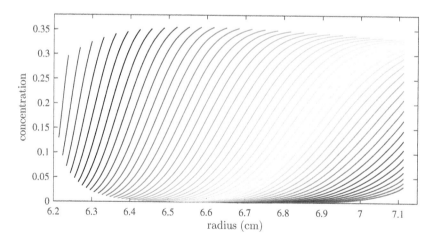

Figure 8.4 Examples for partial boundary modeling. Based on the sedimentation data of the bimodal boundary shown in Fig. 5.8B on p. 107, a selection was made to include only data points with radius values r^* within a range defined for each scan as the range corresponding to s^* from 2 S to 6 S, such as to focus on the data of the 50 kDa, 4 S species underlying the main, slower sedimenting boundary.

In the choice of data limits it is important that the sigmoid shape of the boundary of interest can be recognized. Overly narrow s-ranges will only show central linear slopes, which may lead to parameter correlation since s and D may not be separately identified. Similarly, when fitting the partial-boundary data with a single-species Lamm equation model that includes both TI and RI noise, the molar mass is only poorly determined.[9] The reason for this behavior is that mismatches in the slope of the model can be compensated for by the slope in the TI profile, and the RI parameters can adjust for each scan the absolute signal offsets. (In some cases, the local mismatch of the central slope increases with time, resulting in a curved TI profile.) This correlation can be broken when the RI parameters are fixed *ad hoc*, for example, to values pre-determined from an initial fit with another model. Also it is important in the selection of $s_{PB,min}$ and $s_{PB,max}$ that sufficient overlap of the included data sets from each scan exists.

Partial boundary modeling often simplifies the analysis by restricting the experimental data set. As such it is compatible with all SV models, including the hybrid discrete/continuous distribution. It can be particularly useful in conjunction with explicit discrete Lamm equation models for the sedimentation of reacting systems (to be described in a forthcoming volume), which are outside the distribution framework. It can be freely used in the context of global modeling, with other truncated or complete data sets. However, it should be used sparingly, and, if possible, the explicit incorporation of faster sedimenting minority species into the model *via* sufficiently wide s-ranges in distribution models is preferred.

> Because partial boundary modeling is useful predominantly in models of discrete interacting species, outside the distribution models of SEDFIT, it is implemented only in SEDPHAT. It can be switched on in the experimental parameters box of the SV data type by setting the **partial boundary fitting** flag and entering $s_{PB,min}$ and $s_{PB,max}$.

8.1.5 Selecting the Baseline Noise Models

The noise models are usually straightforward on the basis of the properties of the different optical signals discussed in Part I, Section 4.3. Briefly, the consideration of both RI and TI noise offsets are required for modeling interference optical data, but no RI contributions are usually present in absorbance optical data (unless a buffer component is used that changes absorption with oxidation state, or when working with time-varying speed data sets). TI contributions are always significant in absorbance data, too, but they only produce series of localized blips arising from

---

[9]Partial boundary modeling introduces a mild non-linearity in the TI noise computation due to the only partial overlap of experimental data [38]. This requires an iterative approach. To speed up the convergence of the systematic noise offsets, it is useful to subtract a first estimate of the systematic baseline offset from the data prior to the partial boundary modeling, in order to reduce the magnitude of systematic noise parameters.

small scratches in the window. For absorbance data, therefore, there is the option to leave TI noise out of the analysis, at least at a first stage. Fluorescence optical data should ideally not need either RI or TI noise, although sometimes radially non-constant baselines are encountered that must be modeled using TI noise.

Generally, it is a good idea to subtract the TI and RI estimates from the raw data after a fit. This does not change the statistical properties or the information content of the data, as long as the same noise model is maintained in further analyses. Subtracting the values even from a very preliminary fit can allow us to visually inspect the data and the fit in much more detail. In some circumstances, it can be useful to restore the original data offsets, in particular the TI noise, for example, to better examine the meniscus and bottom regions, or to examine the TI noise profile.

In the vast majority of cases, the RI and TI contributions present no further complications. However, the shape of the best-fit TI profile should be inspected along with the boundary model (below).

In SEDFIT the noise parameters are specified through checkboxes in the model parameter boxes. In SEDPHAT, because noise parameters belong to local properties in the context of global fitting, they are specified in the experimental parameter box of the SV data type.

8.1.6   Inspection of the Fit

For a meaningful analysis, it is critically important to inspect the quality of fit between the sedimentation model and the experimental raw data. The first, and most direct way to do this is a visual inspection of where the model matches the data well and where it doesn't. Although not quantitative in nature, it is very intuitive and can frequently reveal important insights in the origins of misfit, and thereby aid in the improvement of the model. The visual inspection of the fit is also aided by the plot of residuals. Besides the standard overlay plot, very useful tools for detecting systematicity are the residuals bitmap [67] and residuals histogram [68], as reviewed in detail in Part I, Section 4.3.5. An example for a good *vs* bad fit is Fig. 4.29 in Part I.

It is useful to graphically zoom in, for example, to the region of the meniscus, in order to visually assess whether the model overlays the data points or systematically lags behind. The latter may be associated with a variety of potential problems, including convection, lack of meniscus refinement during the fit, incomplete boundary models, and non-ideality. Another region of interest is the solvent plateau: any deviation from a constant signal can display experimental instabilities, such as baseline fluctuations, or slowly sedimenting material. If systematic residuals are found co-migrating with the boundary, it is instructive to zoom in on the sedimentation boundaries in the raw data plot. For example, if the best-fit model after refinement of diffusion parameters is initially too steep and later too broad, then the experimental boundary may be subject to self-sharpening from repulsive non-ideality at high concentrations (Part I, Section 2.2.2.2).[10] The opposite pattern may indicate excess heterogeneity or diffusion unaccounted for in the model. The solution plateau ahead of the boundaries should be flat parallel lines in the absence of broad distributions of large particles. Systematic deviations in this region can indicate either buffer mismatch, aggregates unaccounted for, or effects of static or dynamic density gradients. Finally, the region in the onset of back-diffusion can reveal optical artifacts close to the bottom of the cell, and guide the refined selection of fitting limits. For example, upon closer inspection, systematic discontinuities in the data or residuals plot may be discerned close to the meniscus or bottom that are not a form of noise, and cannot possibly arise from monotonically increasing concentration gradients with any sedimentation process, thus must be optical artifacts that should be excluded from the fit.

Importantly, it is frequently possible to relate the details of the derived distribution immediately to features in the raw data. For example, the persistent slow depletion of material at low radii in the data of Fig. 5.15 on p. 126 directly reveals to the trained eye the presence of low-molecular weight species continuing to sediment throughout most of the recorded sedimentation process. Similarly, in Fig. 5.14 the stretched appearance of the leading boundary portions can convince us very directly of the presence of oligomeric species. The relationship between $c(s)$ distribution and data features can be graphically emphasized in **SEDFIT** using the integration tool.

> To facilitate establishing the relationship between populations of sedimenting species predicted in the $c(s)$ analysis and corresponding features of the sedimentation boundaries, the integration tool of **SEDFIT** generates a graphical highlight of data points contributing to the integrated s-region. Boundaries are colored gray if they do not report on the s-range, and red if they do. The red shaded transition zone reflects an approximation of the extent of diffusional spread of particles with s-values in the region of interest.

If the model allows for a TI noise profile to be calculated, in a second step of the inspection the original TI noise of the data should be added back to restore the raw data, such that the shape of the TI noise profile can be examined. As discussed in detail in Part I, Section 4.1.4, the TI noise will depend on the data acquisition system in characteristic ways. Localized blips, for example, are usually plausible results of window imperfections and correlate little with sedimentation parameters of interest in the model. However, if the best-fit TI noise exhibits unusual features this can be a flag of an incomplete sedimentation model. Especially gently curving lines extending across the solution column with increasing positive or negative slopes

---

[10]This is often accompanied by larger than expected best-fit $f_{r,w}$-values.

toward the bottom of the cell are generally suspect.[11] In some cases the inclusion of additional data (earlier and later scans) helps to constrain the sedimentation model and improve TI profile.

A quantitative measure for the goodness-of-fit is the rmsd. The absolute value of the rmsd has to be interpreted in the context of the statistical noise of data acquisition. The latter is dependent on many factors, as outlined in Part I, Section 4.3.5, and generally not precisely known. More often, relative changes in the rmsd are used as a criterion to accept or reject a fit in F-statistics [71, 72, 132].

Another useful number is the ratio of total signal to rmsd. For a perfect fit, this value will depend strongly on the noise of the data and the optical system used, and it may reach 100–1000:1 for excellent data and fits. If the total signal modeled is extremely large to begin with, then an rmsd value somewhat exceeding the statistical error and some systematicity in the residuals may be acceptable, if the ratio of total signal to rmsd is still very high. For example, when modeling interference optical data that have total loading concentrations in excess of 10 fringes, it would be exceptional to achieve fits with rmsd better than a few hundreds of a fringe; even though this is larger than the statistical noise of data acquisition, signal/rmsd ratios of 1000:1 are at the limit where a variety of common idealizing assumptions on the sedimentation process and/or the data acquisition break down.

8.2 FINE-TUNING THE DISTRIBUTION ANALYSIS FOR NON-INTERACTING SYSTEMS

Because of the ubiquitous presence of impurities in the experimental samples, the distribution models are the most common types of models for the study of non-interacting systems. In this section, we discuss practical issues concerning how to finely adjust the settings of the distribution models, and possibilities for increasing the information content for the species of interest, using the tools introduced above. Common questions, such as the determination of s_w-values, the frictional ratio, and the precision of these quantities, are also addressed. Finally, an overview is provided of which model to choose, depending on the type of sample, to best determine the molar mass of a sedimenting species.

8.2.1 The Lower Limit s_{min}

In sedimentation experiments, the smallest s-value for the distribution analysis may be zero. However, this value describes a species that does not sediment at all. From the experiment we can hope to gain information only on sedimenting or floating species. Computationally this is expressed in the fact that the concentration parameter for this species is completely correlated with the baseline value. When using maximum entropy regularization, this correlation usually causes low baseline values

---

[11]SEDFIT and SEDPHAT calculate the TI noise profile across the entire data set, but the sections outside the fitting range, indicated by the two green lines, is irrelevant.

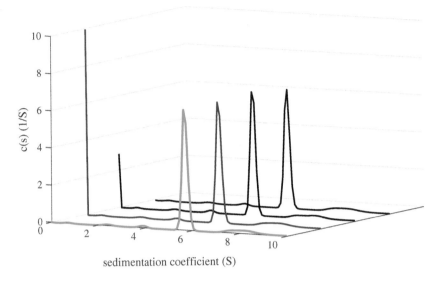

Figure 8.5 Using different values for s_{min} in the $c(s)$ analysis with maximum entropy regularization applied to the sedimentation data of an IgG sample shown in Fig. 5.2 on p. 95. There is no signal from sedimenting buffer salts, since they are optically well matched with the reference solution. Therefore, the distribution at $c(s_{min})$ should vanish. All distributions shown lead to a statistically indistinguishable value for the rmsd. s_{min} values are 0.1 (blue), 0.01 (black), and 0 S (red) without Bayesian modification, and 0 S (green) with Bayesian modification.

and high values for $c(s = 0)$ [209]. Fortunately, one can resolve this correlation elegantly with the Bayesian approach setting the prior expectation for the population of the species with lowest s-value to be slightly lower than that of all other species. As a consequence, the distribution values $c(s_{min})$ will be populated only if there is information in the data on these species. This allows us to work conveniently with very small s-values, including $s = 0$ S (Figs. 8.5 and 8.6). Since buffer salts frequently sediment with an s-value of a few tenths of one Svedberg, and they often do contribute signals, the s_{min}-value for high-speed SV experiments should usually be at least as small as 0.2–0.5 S to ensure a good fit.[12]

Suppression of the population of $c(s_{min})$, if possible without statistically significantly penalizing the quality of fit, is the default in SEDFIT. It can be toggled off and on in the function Options ▷ Size Distribution Options ▷ suppress baseline correlation in Max Ent.

The most critical criterion for the choice of s_{min} is that it does not constrain

---

[12]This is assuming high-speed centrifugation, at rotor speeds above 20,000 or 30,000 rpm. When studying very large particles at low rotor speeds, the minimum s-value would be much higher. For example, in the characterization of the *Spiroplasma* sample of Fig. 4.2 on p. 80, the rotor speed was 3000 rpm and the s_{min}-value was 100 S.

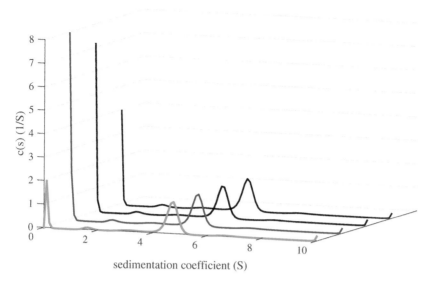

Figure 8.6 In contrast to Fig. 8.5, this plot shows the effect of using different values for s_{min} in the $c(s)$ analysis of an SV data set where buffer salts do significantly contribute to the signal (raw SV data are shown in Fig. 5.15 on p. 126). In this case, using the plain maximum entropy regularization shows a half-peak at $c(s_{min})$ for all values of s_{min} as a result of the effects from both the buffer signal and the baseline correlation. s_{min} values are 0.1 (blue), 0.05 (black), and 0 S (red) without Bayesian modification, and 0 S (green) with Bayesian modification. With the Bayesian modification, a half-peak exists when s_{min} is too large to describe the signal contribution well. At a s_{min}-value of 0 S (green), the distribution correctly shows a full peak for the buffer signal.

the quality of fit. This can be examined from a comparison of the rmsd in a series of analyses with different s_{min}, or it can be assured for the choice $s_{min} = 0$ S. An indicator that the s_{min}-value might be constraining the fit and that it should be lowered is the presence of the half-peak at $c(s_{min})$. However, when not using the Bayesian approach to suppress the baseline correlation, the latter would also result in a half-peak that increases in height with decreasing s_{min}. The typical behavior of $c(s_{min})$ in the presence of signals from small buffer components is illustrated in Fig. 8.6.

As an alternative strategy to cover the description of small buffer components in the distribution, a variation of the $c(s)$ analysis is possible, in which the first s-value, s_1 is treated separately as a non-linear parameter to be optimized in the fit, along with the apparent molar mass value M_1 associated with this s-value (as in Eq. (5.10) on p. 106). As described in Section 5.4.2, an advantage of this continuous $c(s)$ distribution with one discrete component model is that the buffer component will not be included in the weighted average frictional ration $f_{r,w}$, providing a better description of both the buffer sedimentation as well as the particles of interest. By and large, once significant buffer signals have been identified via a standard $c(s)$ analysis, this more detailed model treating the buffer signals as a discrete component is advantageous. A consequence of this approach for the choice of s_{min}

of the continuous segment is that it should be chosen larger distinctly than s_1, such that the distribution value $c(s_{min})$ and the amplitude of the discrete component $c(s_1)$ do not correlate. For example, for buffer salts with the typical $\sim 0.2 - 0.3$ S, s_{min}-values of 1.0 S usually work well.

Another context in which the highest possible s_{min} should be chosen is a logarithmic spacing of the s-grid of the distribution. This can be useful to cover a very wide s-range, but grid points will have the highest density close to s_{min}. With logarithmic spacing of the distribution, s_{min}-values cannot be zero.

For flotation experiments, s_{min} will have a negative sign, and the same considerations apply as described below for s_{max} in the sedimentation experiments.

8.2.2 The Upper Limit s_{max}

Although the time range of the scans and the radial analysis range will set an upper limit to the sedimentation coefficients that can be described (Eq. (3.4) on 62), in practice much lower values for s_{max} frequently are sufficient. Too high values are not problematic, in principle, but can limit the efficiency of the analysis since a larger number of grid points (see below) is required to avoid an overly coarse distribution, and the default distribution display does not represent the region of interest well.

The ultimate criterion for the upper limit of the s-range in the distribution is

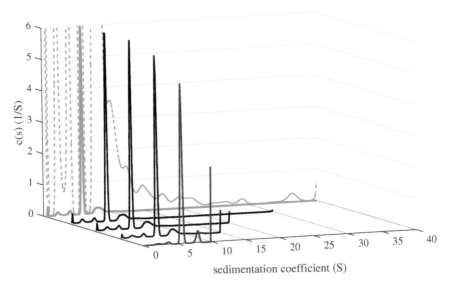

Figure 8.7 Using different values for s_{max} in the $c(s)$ analysis applied to the sedimentation data from the IgG sample. Too low values will not allow the description of all sedimenting species, produce half-peaks at $c(s_{max})$, and result in inferior quality of fit, as visible from the rmsd values and the residuals bitmaps of the fit. Shown are the distribution for values of 10 S (red) leading to an rmsd of 0.0057 fringes, 20 S (black) leading to an rmsd of 0.0052 fringes, 30 S (black) leading to an rmsd of 0.0050 fringes, and 40 S (green) leading to an rmsd value of 0.0050 fringes. In order to highlight the trace populations of species that were excluded from the description at the lower values of s_{max}, the dashed green line shows the same distribution at 100-fold magnification.

that it is not too small to describe all sedimenting species. This can be established, just like in the case of s_{min}, with the help of the visual inspection and the rmsd of the fit. If the s_{max}-value is too low and imposing constraints on the fit, the distribution will exhibit a half-peak at the maximum value (Fig. 8.7). In this case, the s_{max}-value needs to be increased, and in doing so the rmsd of the fit will improve. The adjustment of s_{max} is necessary even if the faster sedimenting species are of no interest in the SV experiment, since the failure of describing all sedimenting species can lead to a bias in the results for all species, including those of the main interest.[13]

When increasing the s_{max}-value, attention has to be paid to the grid spacing of the distribution, so as to not thin out the grid points too much to decrease the quality of the fit. This is discussed next.

8.2.3 Resolution and Spacing

The total number of grid points N needs to be large enough to describe all detail of the distribution resolvable within the given signal/noise ratio. This is the case when it does not constrain the quality of fit. A good starting point that works well for most situations is $N = 100 - 200$, by default uniformly spaced. While the total number of grid points should not be excessively high, because the computational effort increases approximately with N^2, too low values of N will not allow a sufficiently detailed description of the populations of sedimenting particles and increase the rmsd (Fig. 8.8).

Therefore, empirically, the minimal necessary N-value could be assessed by probing whether an increase in N will cause a decrease in the rmsd of the fit. In practice, this test should be carried out after all non-linear parameters, like meniscus and $f_{r,w}$ have been refined to the best possible fit with given N. Whether increasing N will decrease the rmsd will depend on the width of the particle size distribution, the signal/noise ratio of the data, the range of the distribution to be described, as well as the relative position of the s-grid and the peaks in the distribution. For a sufficiently high N the rmsd will not change significantly when doubling the resolution or changing the grid location (Fig. 8.8).

As an alternative, a simple criterion for an adequate N is based on the number of grid points in each peak that are significantly populated. If there are at least two, then any particle sedimenting with an s-value in between these two grid points can be described approximately as a linear combination of the sedimentation patterns

---

[13]There are rare cases when s_{max} determined by this criterion reaches seemingly unreasonably large values. For example, data exhibiting artificially sloping solution plateaus may exhibit this behavior. There are different potential causes for artificially sloping plateaus, including unaccounted for buffer signals in the reference sector in interference data, or radial gradients in signal magnification for fluorescence data, which may be addressed by suitable modification of the model. The exclusion of early scans may help in modeling the plateau region, but in the process diminishes the information on the sedimentation boundaries of interest. Partial boundary modeling (Section 8.1.4.1) can be a useful strategy, as well as the application of a custom s-grid where a regular spacing and range is appended with one or few very large values.

based on the s-values of the neighboring grid points. To ensure a good description essentially independently of the grid location, a sufficient and conservative criterion would be to have three significantly populated points (black dotted line in Fig. 8.8).

Frequently, one chooses first a resolution that is expected not to constrain the fit during the non-linear optimization steps, followed by a final fit at a higher resolution that provides a smoother, visually more attractive result. This final step at a higher resolution can also be useful to confirm that the resolution is sufficiently high to describe all features of the sedimentation coefficient distribution, and that N is sufficiently high such that the quality of the fit is indeed independent of N.

In order to limit the total number of grid points to computationally convenient levels while maintaining a sufficiently close spacing in the regions of the peaks of interest, it is sometimes useful to deviate from the uniform spacing of the grid. In principle, one can apply any distribution of grid points that provides sufficient density. One particular choice of grid spacing that allows us to maintain a high density of points in regions of low s-values, but to also cover a wide range of larger s-values at coarser intervals, for example, to describe aggregate species that are not of primary interest, is the logarithmic spacing. As mentioned above, the combination with a separate discrete component describing the buffer signals is advantageous.

> The resolution and the spacing of grid points is controlled in SEDFIT from within the model menus. N is entered in the top left field resolution. The default setting will result in a linear spacing between s_{min} and s_{max}. If the checkbox log spaced s grid is checked, a logarithmic placement of s-grid points will be used. The checkbox s-grid from file allows complete user-control over all s-grid points, as specified in the ASCII file sdist.* with the same extension as the data files loaded. This file will will be automatically generated by SEDFIT even if s-grid from file is unchecked. The easiest approach is to let SEDFIT generate this file in a preliminary analysis, and then to use a text editor to modify the file, for example, by inserting new values or appending numbers to increase the range.

8.2.4   Regularization in Practice

The concept of regularization was reviewed in Section 3.3. It is very important to keep in mind that regularization produces one particular distribution from all possible, statistically indistinguishable ones. The choice of the type and magnitude of regularization introduces a slight, but statistically insignificant bias to select the most parsimonious distribution. This is essential because the absolute best-fit solution can (and frequently will) be dominated by artificial spikes that are introduced from noise amplification, presenting unreliably detailed results [56].

The need for regularization can be eliminated by using a very coarse grid — in fact, this is a form of regularization, although a poor one.[14] The most commonly

---

[14]If the resolution is very low, then there is no danger of obtaining unwarranted levels of detail.

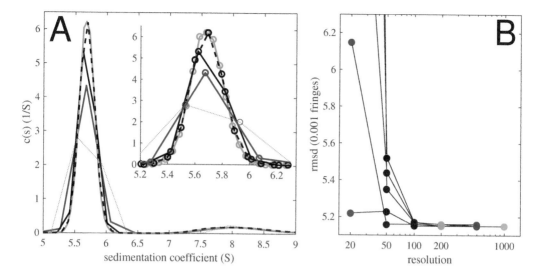

Figure 8.8 Effect of different resolution settings on the peak shapes (A) and the rmsd of the fit (B) for the IgG data. The average spacing of grid points $\Delta s = (s_{max} - s_{min})/N$ should be maintained fine enough to provide three significantly populated points in the peaks of interest. *Panel A*: The $c(s)$ distributions shown here are for Δs values of 0.4 (red and red dotted), 0.2 (blue), 0.1 (black short dashed), and 0.05 S (green). The inset highlights the main peak and shows the grid points as symbols. For the coarse grid (red solid line), there is only one grid point in the peak. If the same grid is shifted, maintaining the same resolution, two points appear in the peak (thin red dotted line), but at a highly increased rmsd of the fit, and with side effects on the secondary peak at 8 S. Acceptable fits appear at resolutions of 0.1 S (black dotted line) or higher. *Panel B*: The rmsd as a function of N covering a range from $s_{min} = 0.5$ S to $s_{max} = 20$ S. The resulting grid spacing is indicated by colors equivalent to the curves in A. For each resolution setting, several values are shown, resulting from the successive shift of s_{min} by 0.1 S. This shift should not significantly alter the rmsd for adequate N, but can have a large influence on too low N-values.

applied regularization methods in biophysical techniques are Tikhonov–Phillips regularization (Eq. (3.8)) and maximum entropy regularization (Eq. (3.9)), and there is a sizeable literature on the merits of different approaches, for example, in dynamic light scattering [210, 211]. The selection of which method to use depends on what is known about the sedimenting particles under study, in terms of the shape of the expected distribution.

The Tikhonov–Phillips regularization is based on the strategy of suppressing unnecessary peaks by minimizing the total curvature of the distribution. This method has been applied famously by Provencher in the program CONTIN [56, 122], widely used, for example, in the field of dynamic light scattering. This regularization method performs particularly well for cases where we expect a relatively broad distribution of species, such as in the study of synthetic polymers. Another field of favorable use for this approach is the ls-$g^*(s)$ distribution, since we expect diffusion

---

Unfortunately, it is unsatisfactory both with regard to the quality of fit, and with regard to the ability to extract precise values for the parameters of interest. This approach should not be used.

to broaden the apparent sedimentation coefficient distribution. The drawback of the Tikhonov–Phillips method is that it does not perform as well for distributions with sharp peaks.

The maximum entropy method is fundamentally very appealing because of its basis in information theory [126, 212]. Pragmatically, it generally shows superior performance for cases where we expect a few discrete peaks. This is the case for most studies of biological macromolecules. Unless they exhibit, for example, a very broad range of glycosylation, and unless we are faced with the effectively

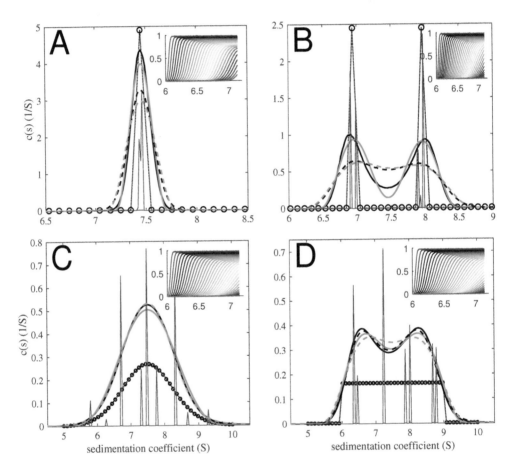

Figure 8.9 Effect of regularization on different types of distribution. Sedimentation data were simulated for the distributions shown as black dotted lines and circles (reduced in scale by a factor 2): a single peak (*A*), a double peak (*B*), a Gaussian distribution (*C*), and a rectangular distribution (*D*). For the simulations, a frictional ratio of 1.2 is assumed, a solution column from 6.0 to 7.2 cm, and one unit total concentration with noise of 0.005 signal units. The insets show the shape of the resulting sedimentation boundaries. For analysis of each data set different types of regularization are used: no regularization (red lines, reduced in scale by a factor of 10), maximum entropy regularization at a confidence level of $P = 0.68$ (blue solid) and 0.95 (blue dotted line), and Tikhonov–Phillips regularization at a confidence level of $P = 0.68$ (green solid line) and 0.95 (green dotted line).

sedimenting particle being a dynamically linked complex (i.e., the presence of chemical reactions on the time-scale of the SV experiment), proteins and nucleic acids can usually be expected to produce sharp peaks in the sedimentation coefficient distribution. Where maximum entropy does not give satisfactory results are broadly distributed peaks [135]. In these cases, we can frequently observe some degree of oscillations. Although rarely observed, another artifact one should be aware of is a possible slight shift of peaks if they are in close proximity (for example, Fig. 8.9B). Another problematic scenario can be the characterization of peaks on top of, or in combination with, a broad elevated background [131], but this seems to rarely be encountered in particle size distributions or sedimentation coefficient distributions.

Fig. 8.9 shows the performance of the different types of regularization. Since regularization always favors broad peaks (unless enhanced with Bayesian prior probabilities), it can be non-trivial to distinguish a single discrete peak (A) from two closely spaced peaks (B), a true Gaussian distribution (C), or a rectangular distribution over a narrow s-range (D). Fortunately, on the basis of prior knowledge of the samples, the distinction can usually be made beforehand. How broad the peaks are will depend on the level of regularization. Usually a P-value of 0.68, corresponding to the confidence level of one standard deviation is sufficient. For the same model, the peak width will also depend on the noise of the data acquisition, the loading signal, the number of scans included, the solution column length, the rotor speed, the systematic noise model, and all other factors that could impact on the information content of the data. Therefore, it is often very difficult to directly interpret or compare the width of the peaks from the distributions without Bayesian methods.[15]

In Fig. 8.9, we can see the maximum entropy method producing sharper peaks than the Tikhonov–Phillips method for the discrete peaks, but exhibiting more oscillations for the uniform broad distribution. Both methods perform quite well for a Gaussian distribution. It should be noted that the attempt to use no regularization fails throughout, and spectacularly so when the distribution is broad.

Since all distributions are statistically equivalent interpretations of the data, it can be useful to calculate them both, and on different confidence levels, in order to obtain a measure for the flexibility in the result. To further probe the set of possible solutions, and to carry out a more specific analysis in light of other data or knowledge about the sample, the Bayesian approach can be used.

8.2.5 Enhancing the Resolution with Bayesian Prior

Regularization with Bayesian prior, as introduced in Section 5.7, is a tool to probe the space of distributions that are statistically consistent with given data. The appeal of this refinement is that it cannot produce distributions that contradict

---

[15]In particular, it is a misconception to attribute the width of the $c(s)$ distribution to a molar mass, such as in the width of a $g^*(s^*)$ peak in the presence of diffusion: In $c(s)$, diffusion and molar mass information is already deconvoluted and separated from the sedimentation coefficient distribution.

the data or lead to a worse fit. We have already listed a few of the applications in Section 5.7. Possible types of prior knowledge include:

known species s-values of a species known to be present in the sample from independent experiments.

dilution series $c(s)$ obtained a high concentration can be used as a prior hypothesis for data acquired at low concentration, as illustrated in Fig. 5.13 (p. 123): This can resolve with much higher sensitivity the question whether or not there is a concentration-dependent shift in the sedimentation coefficient distribution.

test of interaction The prior may embed the s-values of components of an interacting system studied individually side-by-side with the mixture in the same run. The resulting $c^{(p)}(s)$ distribution of the mixture should exhibit additional peaks (or peak shifts) from interactions with better resolution, if complexes are formed. Here, the prior $p(s)$ may be assembled manually from Gaussians or δ-functions placed at the specific, known s-values.

mono-dispersity The s-value of the peak from the initial $c(s)$ analysis of the same data can be turned into a δ-function prior in the $c^{(p\delta)}(s)$ approach. This can highlight, for example, whether or not any given peak describes a single species (and the finite width is solely a result of the regularization in the presence of finite signal/noise ratio), as in Fig. 5.14 on p. 124, or if there is heterogeneity of sedimenting species as in Fig. 5.15. In the latter case, the resulting $c^{(p\delta)}(s)$ distribution should be examined for coverage of the material in the prior peaks, and the presence of unexpected side lobes.

multi-component analysis The $c(s)$ distribution determined from data of the same cell, but using a different optical system may be used as prior: In this case, the question is addressed whether or not we are detecting the same distribution with different optical systems. The Bayesian analysis here eliminates the problem of dissimilar signal/noise ratios limiting a direct comparison. The prior $p(s)$ is taken directly as the $c(s)$ distribution of the signal with the highest resolution.

undisturbed boundary Sedimentation theory for two-component heterogeneous interacting systems generally predicts the presence of a bimodal boundary with an undisturbed boundary coinciding in s-value with one of the free species [75, 213]. The latter can be measured separately, and applied as a δ-function prior in the analysis of the interacting system, so as to enhance the resolution between reaction boundary and undisturbed boundary, for example, for isotherm analysis [70].

enhanced trace component quantitation When using SV to determine the amount of trace aggregate species in highly pure protein samples (such as for quality control in biotechnology products), the use of a Bayesian prior consisting of a δ-function for the main species in combination with a Gaussian for the trace species can improve the precision of the estimated loading concentration [59, 170]. This addresses the limitation that the preference for parsimonious distributions in the standard $c(s)$ distribution will tend to attribute some signal to material close to the s-value of the main peak in order to achieve a smoother peak shape. This lowers (within the statistical limits) the amount of trace aggregates, typically by $\sim$0.1–0.2%. At the same time, it can be difficult to localize the s-value of the trace material at low abundance in the naïve $c(s)$ approach. The Bayesian $c^{(p)}(s)$ analysis addresses both problems [59].

probing variability It is possible to use the Bayesian approach with entirely hypothesized shapes for the prior $p(s)$. Since the resulting $c^{(p)}(s)$ distribution is guaranteed to give the same quality of fit as the $c(s)$ distribution with standard regularization (and statistically indistinguishable to the overall strictly best non-regularized least-squares fit), in the absence of additional prior knowledge, any of the distributions are possible.[16] This can be used to actively explore the confidence region of the analysis in the distribution space, by asking the question "to what extent can the distribution possibly adopt a certain shape?" This can be very useful in telling us how much to rely on certain features of the analysis result.

suppressing baseline correlations As outlined above, the Bayesian approach can eliminate a correlation between very small species and baseline signals by assigning $p(s_{\min})$ a lower value than the rest of the distribution.[17]

With any of these variants, the effects of modified regularization are restricted to the concentration parameters across a distribution for a given set of non-linear parameters of a model. Therefore, non-linear parameters, such as meniscus and average frictional ratio, can be optimized independently before the refinement of regularization. They do not have to be re-fit for the application of Bayesian priors.

8.2.6 Accounting for Buffer Signals

The effect of buffer contributions to the measured signal evolution as a result of chemical or volume mismatch between sample and reference solution was described in Part I, Section 4.3.4. Details of the application of such buffer models are described in [47].

---

[16]However, when reporting the results, as a matter of transparency of the analysis, it should be stated which choice was made.

[17]Rather than enhancing the probability of species at a certain s-value, lowering it makes the distribution try to avoid such species, and to populate it only when there is significant information in this regard in the data such that positive contributions at this s-value are essential to avoid decreasing the quality of fit.

In many cases, a buffer signal can be accounted for by using the distribution model variants with an extra discrete component. This frequently works well if either the mismatch is small, or if the buffer component is absent from the reference sector. More difficult, but occasionally necessary when interpreting the interference optical signals, is the explicit modeling of buffer sedimentation signals in both the sample and the reference sector. This amounts to a positive and a negative small-molecule sedimentation pattern, with separate menisci and potentially separate s and D-values. This leads to several extra parameters that can jointly create a difficult error surface with local minima, making the optimization difficult. However, once the correct parameters have been found, usually a very significant improvement in the fit quality can be achieved.

To arrive at good buffer mismatch models, a useful strategy is a hierarchical sequence of constrained parameter subsets for which reasonable estimates can be made, while optimizing others that are less well known. For example, at first we can graphically determine the extent of geometric mismatch between the menisci in the reference and sample sector and fix this parameter (essentially making the assumption that the meniscus artifact shape is the same in sample and reference). Similarly, we may be able to fix the total signal of the buffer to an approximately expected range, if experience with the co-solutes in the buffer is available (such as the knowledge that NaCl produces ~17 fringes loading signal for each 100 mM).[18] This will break the correlation between the loading concentration and the sedimentation and diffusion coefficients. Starting values for buffer salt sedimentation coefficients may be ~0.2–0.5 S, and apparent molar mass values on the protein $\bar{v}$-scale in the range of a few hundred Dalton.

Depending on the particular experiment, if an imperfect chemical match of the buffer is present, it would be reasonable to use the ratio of buffer loading signals in the sample and reference sector as a floating parameter early on. Otherwise, if the mismatch is expected to be in loading volumes only, the signal ratio in the sample and reference sector can be constrained to unity. In our experience, the consideration of different sedimentation and diffusion coefficients in the sample and reference sector buffers is the most difficult case. Although careful loading of the centrifugal cells can ordinarily avoid this problem, it may not always be possible to control this factor for the experimentalist conducting the SV run, or it may for various experimental reasons not be possible to achieve even a basic chemical match. In these cases, the ability to attribute different s-values and M_{app}-values to the sample and reference sector in the model can be essential.

Once the parameters without pre-conceived expectation values have converged, along with the absolute meniscus position, which should be refined early on, those parameters that were initially fixed should be refined, one-by-one, in a series of fits refining the analysis. Also among the parameters that should be refined, finally, is

---

[18]This knowledge can also be extracted easily from synthetic boundary experiments conducted to study the buffer, with water in the reference sector.

the bottom of the solution column, because it will have a significant influence on the small molecule redistribution during sedimentation.

8.2.7 Accounting for Density Gradients from Co-Solute Sedimentation

Different from the preceding section, co-solutes may not contribute to the signal but dynamically create a significant density gradient during the sedimentation process of interest. The experimental background was reviewed in Part I, Section 2.3.3, and the phenomenology of the resulting sedimentation pattern was discussed above in Section 2.2.5 on p. 44. Here we discuss aspects of the practical application of this model.

The first and most important question is whether it is necessary to account for dynamic density gradients at all. To judge this it is necessary, first, to determine the relative co-solute concentration gradients along the solution column during the experiment, and second, to assess whether these will significantly alter the solvent density and viscosity.[19]

The first goal therefore is to predict co-solute sedimentation and diffusion coefficients. In principle this can be accomplished in a separate sedimentation experiment at high rotor speeds using the interference optical detection, placing buffer with co-solute in the sample sector and buffer without the co-solute (or water) into the reference sector. A single-species fit to the resulting small-molecule sedimentation pattern, which will resemble Fig. 2.5A, should reveal s and D. Unfortunately, for sedimentation patterns with such low sedimentation coefficient, a significant correlation can occur between the total loading signal c_0 and the transport parameters s and D, rendering the latter often unreliable in a standard SV experiment. However, the correlation can be broken if the loading signal c_0 of the given co-solute concentration, in units of fringes, can be determined independently. This is easily possible using a synthetic boundary experiment, as schematically shown in Fig. 2.10 on p. 40 (Section 2.2.3), overlaying the co-solute in the sample sector with solvent lacking co-solute. The experimental setup is discussed in Part I, Section 5.2.1.2. Either the entire evolution of signal can be modeled with a single-species model with synthetic boundary initial condition, or the signal amplitude of the initial steep boundary can simply be visually estimated after empirical RI noise correction.[20] The resulting value can be inserted back into the analysis of the standard SV experiment for co-solute sedimentation.

The next step is the determination of the maximum relative concentration increase and depletion during the SV experiment. Knowing the loading signal c_0, we can determine the maximal signal difference from bottom to meniscus, Δc_{max} occurring at the end of the run or at the last time point of interest, by simulating run

---

[19]Gradients in hydration effects on the macromolecular apparent partial-specific volume should be negligible at the experimentally achievable co-solute concentration gradients.

[20]A function to vertically align interference scans over a user-selected radial region is available in the SEDFIT menu Options ▷ Loading Options and Tools ▷ Eliminate Jitter.

conditions of the original experiment with the macromolecule. The ratio $\Delta c_{max}/c_0$ equals the maximum relative difference in molar concentration during the SV experiment.

The second goal is now easy to achieve, simply by determining how much the sample density and viscosity changes as a result of the molar co-solute concentration differences, and whether ensuing changes in the s-value correction Eq. (1.22) are negligible or need to be accounted for at the desired precision of the analysis. The dependence of solvent density and viscosity co-solute concentration can be predicted or measured, using any of the techniques discussed in Part I, Section 3.2.

Accounting for the dynamic density gradient is now straightforward, using the sedimentation parameters and density and viscosity increments established so far. To this end, SEDFIT will carry out a coupled sedimentation process and locally update viscosity and density data at all time steps to correct for macromolecular migration, using computational procedures described in [18].

The Options ▷ Inhomogeneous Solvent model in SEDFIT will require ASCII text files density.dat and viscosity.dat, using the form $\rho(c) = a_1 + a_2 \times 10^{-3}\sqrt{c} + a_3 \times 10^{-2}c + a_4 \times 10^{-3}c^2 + a_5 \times 10^{-4}c^3 + a_6 \times 10^{-6}c^4$ (with ρ in mg/ml and c in molar units) and analogous for the relative viscosity (without the last term). The coefficients may be taken from the software SEDNTERP. However, fewer coefficients can be used. The files will need the coefficients, and the multiplication with the indicated powers of 10 will be done automatically in SEDFIT. If other buffer constituents or temperatures are present that change the buffer density and viscosity above that of water at standard conditions (even at zero concentrations of the co-solute generating the density gradient), these changes can be accounted for as increments added to the coefficient a_1.

Also required are sedimentation and diffusion coefficients of the co-solute under experimental conditions, which can be determined best in separate experiments using interference optical detection. Due to the high co-solute concentrations, it may seem desirable to account for concentration dependence of s and D, but to the extent that only empirical values are necessary over a fairly limited concentration range, this should generally not be necessary for most co-solutes. As a control for proper calculation, the co-solute concentration gradients can be displayed graphically superimposed on the experimental sedimentation boundaries.

When the Inhomogeneous Solvent model is switched on, all sedimentation coefficients calculated are corrected to standard conditions (deviating from the general rule that sedimentation coefficients are uncorrected experimental conditions).

8.3 DETERMINATION OF COMMON SEDIMENTATION PARAMETERS

8.3.1 The Weighted-Average Sedimentation Coefficient s_w

After the sedimentation coefficient distribution is satisfactorily fitted to experimental data, a survey of the resulting peaks leads to the identification of the species of interest. s_w is then determined by integration following Eq. (3.13). Even if the distribution peak arises from the sedimentation of a single species, provided the analysis results in a good fit of the data, this approach yields the best possible determination of its s-value. Due to the close relationship between the transport method and the integration of a distribution, as described in Section 3.4, the so determined s_w-value rests entirely on the fit to the data matching the change in area of the boundary, independent of the physical motivation of the sedimentation model. Thus, the s_w-value determined in this way is applicable for sedimentation analysis beyond non-interacting systems.

Integration of distributions is carried out in the same way in SEDFIT and SEDPHAT. First, integration is initiated by either pressing the integrate button in the distribution window, using the keyboard shortcut CTRL-I, or invoking the SEDFIT function Options ▷ Size Distribution Options ▷ integrate distribution. This turns the mouse arrow into a cross-hair, waiting for the user to drag — while the right mouse button is kept down — a rectangle in the distribution plot across the area to be integrated, after which the mouse turns back into the arrow. In order to abort the integration mode, the ESC key can be pressed. Otherwise, the release of the right mouse button after drawing the rectangle will cause the integration calculations to initiate. In two-dimensional size-and-shape distributions, this should be done in the "native" display of $c(s, f_r)$ and the rectangle will determine the integration limits for s and f_r, respectively. In one-dimensional sedimentation coefficient distributions, such as $c(s)$, only the left and right limits of the drawn rectangle are relevant. Integration is carried out according to Eq. (3.13) and the results are displayed as text in a message box.

If numerically precise integration limits are required and/or the integration is to be carried out repeatedly across standard sedimentation coefficient limits, the function Options ▷ Size Distribution Options ▷ use c(s) integration ranges from file can be used. This substitutes the graphical input with entries from an ASCII text file containing one or more rows, each containing two numbers specifying a lower and upper s-value of an integration range.

If multiple identical samples are studied side-by-side in the same SV run, and data are acquired and analyzed according to the principles outlined here, typical statistical errors in the s-values of the dominant species are in the 0.01 S range, and even better reproducibility may be achieved by very careful experimentation. The accuracy of the s-value depends critically also on the correct instrument calibration, as discussed in Part I, Chapter 6.

An important question is how we can estimate the precision of s_w from the data analysis. Clearly very noisy data will have a larger statistical error than those from a perfect fit at high signal/noise ratio. Unfortunately, since s_w is a quantity derived by integration of a distribution, it is not easily possible to use the standard approach of probing the shape of the error surface projections to search for the $P = 0.68$ contour by F-statistics [132]. However, a Monte-Carlo approach is possible, where a large number of in silico generated data sets with the same noise structure but different draw of statistical noise are analyzed. In fact, even though many hundred or even thousands of iterations are required for reliable results, this can be accomplished rather quickly because many of the computational steps do not have to be repeated in each iteration [137] (Appendix B).

A Monte-Carlo statistical analysis of the distribution fit to experimental SV data can be carried out in SEDFIT using the functions Statistics ▷ Monte-Carlo for Distributions and ▷ Monte-Carlo for integrated weight-average s-values. The first calculates upper and lower contour limits for the $c(s)$ distribution. However, these are not useful for assessing the uncertainty of an integral quantity, such as the equivalent loading concentration or s_w under a peak, due to the usually strong correlation between distribution values at individual grid points. By contrast, the second function builds up a frequency distribution of the integral over a user-determined range, and reports statistical errors of the integral quantities including s_w.

If the distribution analysis is coupled with the refinement of non-linear parameters, which usually includes the meniscus, and in $c(s)$ the average frictional ratio $f_{r,w}$, in principle a non-linear regression at each iteration would need to be carried out. However, a more effective approach is to determine by conventional F-statistics with error projection the upper and lower confidence limits of the meniscus — this is the parameter usually most correlated with s_w-values — and then to determine for each limiting meniscus value the statistical error of s_w using the Monte-Carlo method. Finally, confidence intervals from both meniscus values should be merged to produce an estimate for the total uncertainty of s_w.

In cases of data with very low signal/noise ratio where the non-linear refinement of the meniscus value is not reasonably possible — or if extreme values are obtained that are clearly larger than the possible range from visual inspection of the artifact region — the graphically determined extreme values for meniscus should be used as the basis for the Monte-Carlo statistics of the distribution for s_w.

If s_w-values from a distribution peak of a given data set are interpreted to be s-values of a physical species, this requires the monodispersity of the corresponding species, as well as reliance on the absence of attractive or repulsive interactions during the sedimentation process, which must be ascertained in a concentration series leading to concentration-independent s_w.

8.3.2 Species Concentrations

A considerable interest in determining relative species concentrations as accurately as possible exists in the pharmaceutical industry, where concentrations of potentially immunogenic oligomers and aggregates of protein therapeutics need to be monitored [214–218].

In practice, the determination of the concentrations associated with a peak in the $c(s)$ distribution follows the same rule as s_w, and since peak integration is part the calculation of s_w the loading concentration of species associated with a certain

s-value range will be displayed alongside s_w. The same Monte-Carlo analysis can be used. The statistical precision that can be achieved for the amplitudes of trace species will depend strongly on the s-values of the species of interest in relation to the dominant species of the sample. For example, quantitation limits are higher for dimeric species sedimenting in the leading edge of the sedimentation boundary of a monomer, as compared to larger oligomers and aggregates that form boundaries that are hydrodynamically completely separate from other species. For the latter scenario, the detection limit can be far lower than the noise in a single data point, since the plateau levels defining boundary heights (i.e., loading concentrations) are usually described by thousands of data points (Fig. 8.10).

As already indicated in Section 5.8 and illustrated in Fig. 8.10, there is a hierarchy of information that can be extracted from SV data. The concentrations are the most noise-robust measurements. It is not unexpected that at very low signal/noise ratios the concentration of a species can already be well defined, while the s-value, or even more the $f_{r,w}$-value, are poorly defined. This should not raise doubts about concentration results.

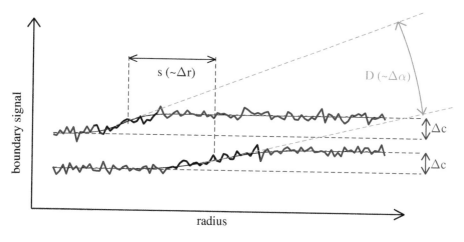

Figure 8.10   Relationship between information content of SV boundaries for the characterization of trace species and corresponding features of experimental data. The cartoon illustrates the different impact of noise on calculated concentrations, s-values, and diffusion coefficients. For clarity we consider just two noisy scans (bold solid lines). The portions of the scans that define the solvent and solution plateaus (magenta) contain a large number of data points, making the determination of the pre- and post-boundary signal levels (Δc, magenta dashed lines) — i.e., the species concentrations — statistically very robust. More uncertain are the exact locations of the boundary midpoints: there are fewer data points (black) and the radial point where the signal level is half the difference between plateaus (vertical blue dashed lines) is obscured by noise. The translation of midpoints with time, Δr, defines the s-value. The most noise-sensitive feature is the slope of the boundary at its midpoint (green dashed lines) and its change with time, $\Delta\alpha$; this contains the information about the diffusional spread (e.g., frictional ratios and molar mass values). Only data with good signal/noise ratio will define the diffusional spread well. By contrast, the concentrations can be measured reliably at a very low signal/noise ratio.

8.3.3 The Frictional Ratio in the $c(s)$ Distribution

The first analysis for most samples is the $c(s)$ distribution with weighted-average frictional ratio. Each clearly visible sedimentation boundary contains information at most on a single $f_{r,w}$-value. This parameter needs to be refined during the $c(s)$ analysis.[21] The first goal is to achieve a good fit, so that the accuracy of the s_w-values from the transport method by peak integration is ensured. While it is good practice to take note of the best-fit value of $f_{r,w}$ and judge whether it has assumed a realistic value, this is not essential for s_w.

Before interpreting the best-fit $f_{r,w}$-value, its statistical significance should be ascertained. The most effective way is an error surface projection such as plotted in Fig. 5.5A (p. 103). It can be assembled by fixing $f_{r,w}$ to values increasingly larger than the best-fit value, and noting the increase in the rmsd after allowing all other non-linear parameters — which are usually only the meniscus — to compensate for the constraint. Then the same procedure is repeated for smaller than optimal $f_{r,w}$-values. On either side, the $f_{r,w}$-values where the rmsd achieves a level prescribed by F-statistics on a given confidence level correspond to one extreme value of the confidence interval of $f_{r,w}$. Poorly defined $f_{r,w}$ values may be encountered, for example, in data sets with low signal/noise ratio, in SV data of very large particles that do not diffuse very much during the short time required to traverse the sample column, or in data with a broad distribution of small species where diffusion correlates with polydispersity. On the other hand, the inclusion of shallow back-diffusion in the analysis, if coupled with optimization of the bottom position, will help determination of $f_{r.w}$. This is due to the molar mass dependence of back-diffusion providing additional constraints for $f_{r.w}$. As illustrated in Fig. 8.10 and discussed in Section 5.8 above, $f_{r,w}$ may or may not be well determined, but uncertainties this parameter should not cast doubts on the validity of other features of the data analysis.

> The rmsd value corresponding to a certain confidence value by F-statistics can be obtained in SEDFIT using the Statistics ▷ calculate variance ratio (F-statistics) function. In SEDPHAT this function is Statistics ▷ Critical chi-square for error surface projections. The search for confidence intervals can be automated in SEDPHAT in different ways.

If the molar mass and partial-specific volume of a macromolecule is known, then the most precise avenue to determine its frictional coefficient is its determination via Eq. (1.32), solely requiring experimental determination of the s-value, and side-stepping statistical errors of $f_{r,w}$ in the data analysis. In this case, errors only

---

[21] Usually $f_{r,w}$ and the meniscus constitute the only two non-linear parameters of the $c(s)$ analysis, which allows the minimization to converge rapidly.

propagate from that of s, and uncertainties in the independently known values of M and $\bar{v}$.

In theory, for non-interacting systems — provided that correct buffer density and viscosity and the correct apparent partial-specific volume are used — $f_{r,w}$ should reflect a translational hydrodynamic frictional ratio, and as such assume values for folded proteins typically between 1.2 and 1.5 (Part I, Section 1.1.2). Obviously, different values can be obtained for particles of unusual shape, though not below 1.0, but such values may also flag deviations from the assumption of an ideal sedimentation process of non-interacting species.

As will become clear in the general description of SV analysis of systems with attractive and repulsive interactions, which will be the topic of a forthcoming volume, the influence of interactions on the observed sedimentation process can be assessed by carrying out experiments in a series of different concentrations, spanning as wide a range as possible. In the absence of interactions, $c(s)$ peaks will broaden as a result of different signal/noise ratios, but not shift. By contrast, in the presence of interactions, a shift in the peak position as well as a concentration dependence of s_w after integration of a single peak or of several peaks will be observed. This is an essential control for any conclusion drawn from the SV data analysis directly on physical properties of particles in the sample.

Much lower than expected $f_{r,w}$-values, sometimes even below 1.0, may be encountered with interacting systems (Part I, Section 2.2.1): As shown in the context of effective particle theory, the apparent diffusion coefficient of the reaction boundary of rapidly associating/dissociating heterogeneous systems can be lower than that of a physical particle [76]. Likewise, the drawn-out boundaries of self-associating systems in rapid oligomeric exchange or with species lifetimes on the time scale of sedimentation can mimic excessive apparent diffusion and lead to best-fit $f_{r,w}$ parameter values that are too small.

Much higher best-fit values of $f_{r,w}$ than expected on the basis of the hydrodynamic contour of the particles are typically associated with boundary sharpening in systems with repulsive hydrodynamic non-ideality (Part I, Section 2.2.2.2). In this case, too, $f_{r,w}$ ceases to have a direct physical interpretation, although extrapolation to infinite dilution can restore physically meaningful values, as shown by Pavlov and co-workers [77, 78, 219, 220].

Even where interactions are absent, the frictional ratio $f_{r,w}$ from $c(s)$ constitutes an average of all species contributing to the SV boundary. As such, it will be an excellent estimate for a species represented in a sole or major $c(s)$ peak, but it may or may not apply to species represented in minor $c(s)$ peaks. For these, it can be regarded as a reasonable guess, in light of the relative lack of information on trace species diffusion coefficients. Of course, any prior knowledge available on the nature of the sample under study should be built into the model to improve the analysis (Section 5.4).

8.3.4 The Molar Mass and Molar Mass Distribution $c(M)$

The question of how to best determine the molar mass of a sample under study is tightly connected with the preceding discussions on the determination of s-values, the interpretation of the diffusional spread of the boundary — in particular, $f_{r,w}$ — and the context of prior knowledge of sample properties. Here we describe and contrast different approaches applicable for different types of samples. Throughout, we assume that the partial-specific volume is known (Part I, Section 3.3), along with solvent properties, and that the instrument is properly calibrated (Part I, Chapter 6).[22] Also, we have to assume that species are non-interacting, as verified by experiments over a wide range of concentrations showing concentration-independent sedimentation properties.

8.3.4.1 Monodisperse Samples

The simplest attempt to determine the molar mass of a particle is fitting the sedimentation boundary data with a single-species Lamm equation solution. As we have seen above in many examples, this is usually not a successful approach, since any heterogeneity in the sedimentation coefficient, even from trace impurities, will strongly affect the boundary broadening and therefore increase the apparent diffusion coefficient and, as a consequence, lower the apparent molar mass. This underestimate is exacerbated with increasing molar mass.[23] For this approach to be successful, it requires an exquisitely pure sample, which is in practice extremely rare (despite best practices of purification and apparent homogeneity when tested by many other — less sensitive — methods). A critical inspection of the residuals of a fit comprising the entire sedimentation process will generally show whether or not the single-species Lamm equation model is appropriate. If the residuals to the full data set do not show any systematic deviations, then the single-species Lamm equation model is indeed appropriate, and in that case the best-fit molar mass is

---

[22]However, the solvent viscosity does not need to be known for the determination solely of the buoyant molar mass, since its effects on s and D are equal and cancel out. However, it will be important for conclusions on molar mass drawn on the basis of observed s-values alone, such as the application of Stokes' law to estimate the minimal molar mass of a particle associated with a certain $c(s)$ peak.

[23]This can be shown in a different way with a series of computer simulations of SV data for mixtures of a main species in the presence of subpopulations with 0.9- and 1.1-fold the molar mass (and 0.93- and 1.065-fold s-value) of the main species. At a fraction of 10% for each minor component, and main species of 10 kDa, 30 kDa, 100 kDa, 300 kDa, or 1000 kDa, a fit with a single discrete species model results in apparent molar mass values that are 99%, 96%, 92%, 78%, and 52% of the average value. If the heterogeneity is more substantial and the trace species are 0.5- and 2-fold the molar mass (and 0.63- and 1.59-fold the s-values) of the main species, then the apparent molar mass values of a single-species fit are 80%–90%, 70%, 51%–83%, 65%, and 73%, respectively (with the ranges of values arising from inclusion or exclusion of back-diffusion or different noise models). In any of these scenarios, the $c(s)$ analysis with refining $f_{r,w}$ yields an apparent molar mass value between 99% and 103% the average.

the best estimate possible from SV. Otherwise, we can expect from this approach only to obtain a lower limit of the true molar mass.[24]

An exception to the widespread problems of single-species Lamm equation fits occurs for small particles (or low rotor speeds) where the sedimentation process is governed by back-diffusion across the entire solution column, and the sedimentation patterns are always of the "approach-to-equilibrium" type (e.g., Fig. 2.5B). Frequently, small molecules, such as peptides, can be modeled very well as single species. This is a reflection of the fact that the much stronger diffusion and slower sedimentation of small particles makes the resolution of heterogeneity more difficult, and simultaneously, that the impact on migration from differential sedimentation is smaller relative to the stronger diffusion. Again, the quality of fit should be the guide for the application of this model.

At the other end of the spectrum, for particles that are so large that there is virtually no diffusion taking place during sedimentation, we cannot hope, of course, to determine the molar mass with any accuracy (see also Section 5.6). Boundary broadening here typically arises mainly from the heterogeneity of particle sizes. It can be instructive, however, to use the single-species model in conjunction with theoretical molar mass values assuming spherical particles of the given density in simulations to visualize the maximal boundary broadening for particles sedimenting at the observed s-values. The comparison of the boundary width of such theoretical Lamm equation solutions with the experimentally observed boundary spread can provide a powerful visualization of sample polydispersity and of the hydrodynamic resolution of SV.

For any particle size, a statistical analysis *via* F-statistics or Monte-Carlo analysis of the confidence interval for the molar mass estimate is critical. This will display the information content of the data used. However, it will not include systematic errors from experimental sources or from incorrect assumptions in the model.

A cautious strategy for samples that are suspected to be mono-disperse is the first analysis with a standard $c(s)$ analysis refining the frictional ratio $f_{r,w}$. If a single peak is observed, is may be followed by $c^{(p\delta)}$ with the Bayesian variant embedding a single-species expectation into the regularization (Fig. 5.14 in Section 5.7). If this yields a single peak that can account for close to 100% of the sedimentation signal, a change of model to a single discrete species is warranted (where integration of $c(s)$ will provide excellent starting estimates for s and M). A comparison of the rmsd of the $c(s)$ model and the discrete species model using F-statistics as a criterion for significance of any differences can validate the single-species model best.

---

[24]Especially at very high concentrations, where the sedimentation process is influenced by hydrodynamic non-ideality from repulsive interactions, too high apparent molar mass values may be possible.

8.3.4.2 Paucidisperse Samples

Several strategies are possible to adapt a discrete single species model to the case of paucidisperse samples, i.e., where the sedimentation data of a discrete species are superimposed by those of distinctly slower or faster sedimenting species.

Restrictions of the data set to eliminate data affected by these extraneous species are more successful for larger species, which sediment faster, than smaller species, which tend to quickly impact the entire radial range due to broad boundaries and back-diffusion. Very large aggregates in a protein sample, for example, may often be excluded by eliminating early scans, by partial-boundary modeling (Fig. 8.4), or by restricting the fitting range to a region close to the meniscus in Archibald-like analyses of suitable data (Fig. 2.8). This path presents the temptation to truncate the data to the point that it will fit even the wrong model. Much more reliable is the opposite strategy — to include all data and extend the model to explicitly include all species in the sample.

It is highly advantageous to proceed, as described above, with an initial standard $c(s)$ analysis, which may then be successively refined *via* probing with $c^{(p\delta)}$ for consistency with expectations. If warranted, it may be extended by hard constraints, such as in the hybrid discrete-continuous $c(s)$ distribution described in Section 5.4.2 (Fig. 5.9). In the $c(s)$ analysis, as long as the species of interest is the origin of most of the signal and reflected in a single dominant peak, the $f_{r,w}$ parameter will refine to a value close to the major species' frictional ration, and the $c(M)$ analogue (or integration of $c(s)$) will provide a good estimate of the molar mass.

It should be noted that the $c(s)$ approach will not result in a reliable estimate of the molar masses of the low-abundance extraneous or trace species — these may or may not have a similar frictional ratio as the main species and their M-value will shift accordingly. Similar is true if the $c(s)$ distribution exhibits two or more major peaks. But in this case, at least if the peaks are well separated, we can use a variation with multiple segments assigning separate $f_{r,w}$-values to each peak.

8.3.4.3 Polydisperse Samples

The study of particles with quasi-continuous molar mass distribution, as well as samples with one or more hydrodynamically un-resolved species require analysis with a continuous molar mass distribution $c(s)/c(M)$ in SV. Except when dealing with small molecules, this is true even when there is a major species that comprises 90% or more of the signal. Since the distribution methods $c(s)$ and $c(M)$ will provide conservative results also in the case of mono-disperse or paucidisperse samples, without the susceptibility to unrecognized sample heterogeneity, they are commonly used as the first method of choice, with the option of further refinement if warranted by the data.

For not too broad mono-modal distributions, the standard $c(s)$ with constant frictional ratio (Section 5.4.1) will provide an excellent first approximation of the average mass and the mass distribution. As described in Section 5.4.9, different $c(s)$

flavors with specific hydrodynamic scale relationships are only needed for broad distributions. In fact, for distributions with a relative polydispersity $\Delta M/M <$ 0.1 the effects of regularization likely have a greater influence on the shape of $c(M)$ than details of the scale relationship.[25] Furthermore, for samples consisting of multi-modal distributions presenting clearly distinguishable boundaries in SV, a segmented $c(s)$ distribution can be applied with separate $f_{r,w}$ values for each boundary, each allowed to refine to different values, thus separately approximating different mass ranges well.

By contrast, polydisperse samples continuously covering a very large mass range at once are where the different flavors of $c(s)$ embedding different hydrodynamic scaling laws (Section 5.4) are important, and knowledge of the scale exponent, structural prior knowledge of the samples, and/or calibration experiments with samples of small polydispersity in different s-ranges are indispensable to determine the molar mass distribution $c(M)$.

Finally, for samples that show significant heterogeneity in buoyant molar mass and in hydrodynamic conformations, the two-dimensional size-and-shape distribution $c(s, f_{r,w})$ is the most appropriate. Starting as usual from the $c(s)$ distribution, it may be tested whether the additional dimension is extracting significantly more information, which will be indicated by a significant improvement of the rmsd. Examples for applications where the size-and-shape distribution can be highly beneficial are samples of nanoparticles with different densities [155], and extended protein assemblies with globular and non-globular species [221].

---

[25] For a specific case, this question can be probed easily using different Bayesian priors.

Numerical Solutions of the Lamm Equation

$\mathbf{L}$ AMM equations are of central importance as master equations for SV. Therefore, techniques for efficient and accurate solutions were intensely studied throughout the history of AUC. Even though all of the numerical procedures are now embedded into computer software, familiarity with the basic ideas is very useful. This will allow critical appraisal of the accuracy of the Lamm equation solutions, and adjustments to the default numerical control parameters to optimize performance and/or navigate computational instabilities.

The radial geometry of sedimentation prohibits a closed-form analytical solution. The challenges for the numerical solution are to achieve a sufficient accuracy over a very large range of concentration values (from $\sim 10^{-3}$ in the solvent plateau to 10^3 or higher at the bottom of the cell), an equally extreme range of gradients, and many orders of magnitude variation of the driving forces for different sedimentation conditions. Differences in numerical efficiency are currently barely noticeable any more under most conditions, but were an additional key consideration throughout the 20th century. Several different algorithms for numerical Lamm equation solutions of non-interacting species have been described [11, 29, 34, 73, 74, 222–226], such that a comprehensive description would be out of the scope of the present section.

The numerical solution of partial differential equations is a highly evolved field of practical mathematics, which we do not attempt to represent. In this light, the simulation of the sedimentation processes appears to be a comparatively simple task, yet it has its idiosyncrasies that are non-trivial and important in practice. Therefore, we will only roughly outline the main ideas shared in the different historical algorithms, in a way for consumption by non-mathematician physical scientists, and keep the focus on the basic principles of the adaptive grid-size finite element approach [34] that is currently the most efficient and flexible and is invoked in SEDFIT for most conditions.

For this discussion we assume a single species exhibiting ideal sedimentation, without attractive or repulsive interactions, and, for clarity of discussion, we exclude flotation, as well as time-variable rotor speeds. It is assumed that the loading concentration is 1.0, and the task is to predict the sedimentation profiles at given rotor speeds, s and D with a precision ϵ better than the statistical precision of data acquisition.

All numerical approaches have in common a division of the radial range from meniscus m to bottom b into N_r radius values r_i, and a division of the time range from 0 to T of desired sedimentation into N_t times t_j. The radial grid and the time steps are usually not simply equidistant, and sometimes correlated. Different algorithms prescribe specific radial grids, time steps, and corresponding changes of the concentrations $\tilde{\chi}_{i,j}$ at these radii and time points to approximate the true solution $\chi(r_{\text{xp}}, t_{\text{scan}})$ at desired experimentally observed radii and scan times sufficiently well. Often, suitable interpolation is required to map the numerical to the experimental grid. The overriding criterion for successful Lamm equation solvers is the accuracy and the careful consideration of how we measure the accuracy can be equally important as the algorithm itself.

A.1 REGIONS OF INTEREST

First, it is very useful to clarify the region where accuracy is required and where it is not. Obviously the region of importance is restricted to the fitted radial range of the solution column, which for experimental reasons usually excludes the steep back-diffusion region (see Section 8.1.4). Whether for any given s- and D-value combination the back-diffusion will reach into the radial fitting range can be easily predicted based on the equilibrium distribution as the limit of maximally extended back-diffusion. Conveniently, the ratio s/D defines the buoyant molar mass, and if the equilibrium distribution contributes by more than a threshold ϵ, e.g., $\epsilon = 10^{-4}$, at the right fitting limit, then back-diffusion needs to be considered and the reflective boundary condition at the bottom of the cell needs to be applied. Otherwise the molecules are never observed under the influence of the bottom, and a permeable boundary condition at the bottom may be used instead to simulate sedimentation in a semi-infinite solution column. This can greatly improve numerical stability simply by avoiding the (unobserved) region of the strongest concentration gradients [61].

> In SEDFIT and SEDPHAT the control box for the numerical Lamm equation solution contains a checkbox shut off back-diffusion determining whether the boundary condition of a semi-infinite solution column is allowed, provided a given threshold is not exceeded even for the projected equilibrium distribution. By default this option is switched on. If the threshold is exceeded, then back-diffusion will be incorporated with standard reflective boundary conditions, irrespective of the control box settings. This condition will be evaluated separately for each species simulated in the course of a distribution analysis.

By contrast, a high accuracy is always needed in the sedimentation boundary.

Obviously, the boundary shape is computationally much harder to calculate than the solution and solvent plateaus. In fact, once the boundary has separated from the meniscus, at a certain distance above (i.e., at smaller radii than) the boundary the concentrations will be negligible, $c(r_l^*) < \epsilon$, and consequently, for all radii $r < r_l^*$, one can safely substitute 0 for all concentrations. Due to sedimentation, this region will grow with time. Likewise, below the boundary the concentration asymptotically approaches the constant solution plateau $c_p(t)$, and if $c_p(t) - c(r_h^*) < \epsilon$ the analytical expression Eq. (2.8) may substitute any computation for radii $r > r_h^*$ if no back-diffusion needs to be considered. This region will shrink with time. In this way, the monotonic decrease and increase of concentration behind and ahead of the sedimentation boundary, respectively, can be used to truncate the radial range over which any numerical computation needs to be carried out. Of course, this saving requires the overhead to update the limits r_l^* and r_h^* at each time step. This approach of dynamically dividing the radial grid points in active and inactive regions, introduced in [34], leads to great improvements in efficiency, and focusses the computational effort on the hardest problem — the accurate calculation of the boundary shape.

The adaptive recognition of plateau regions and the analytical determination of their concentration values can be switched off in the `dynamically truncate grid` field of the Lamm equation box. By default this option is switched on.

For large species with small D the plateau regions will cover the vast majority of radial points. In light of their trivial analytical calculation, it is clear that a definition of the accuracy as the average or rmsd deviation across all radial points from meniscus to bottom, such as $\delta = (N_r N_t)^{-1} \sum_{i,j} \left(\chi(r_i, t_j) - \tilde{\chi}_{i,j} \right)^2$, would allow deviations in the actual boundary shape to be very large despite small δ — simply due to the incorporation of the large number of trivial points.[1] Rather, a more useful criterion for the success of a Lamm equation algorithm is the overall maximum error $\delta = max|\chi(r_i, t_j) - \tilde{\chi}_{i,j}|$, which in the semi-infinite solution column will be governed by the data points in the boundary.

A.2 FINITE DIFFERENCE METHODS

A physically inspired approach to solve the Lamm equation is to understand the radial grid as a stationary division between neighboring solution compartments, or bins (Fig. A.1). In the simplest case, r_i may be equidistant, such as $r_i = m + (i - 1)\Delta r$, where $\Delta r = (b - m)/(N_r - 1)$. The changes in concentration within each

---

[1]In fact, as discussed in detail in [137], this is the case in the ASTFEM Lamm equation algorithm [226], where very coarse radial grids are introduced on the basis of the small overall rmsd, but disregarding very large maximum errors in the boundaries shape exceeding typical experimental noise by more than an order of magnitude.

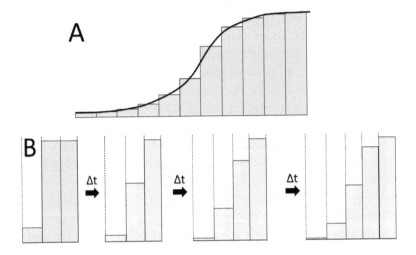

Figure A.1 Schematics of the finite difference approach. *Panel A*: Principle of discretization of the sedimentation boundary (bold black line) into average concentration values within physical volume elements (bins) with width Δr. *Panel B*: Numerical diffusion occurs when the displacement of the boundary is large relative to the bin width. For example, an initially sharp boundary may be well represented by two neighboring bins with dissimilar concentration. After a time step Δt that only translates the boundary by $\Delta r/2$, indicated by light shaded patches in the background, the new average concentration in each bin drops to the average of its previous value and its left neighbor. This averaging happens in each time step and behaves like physical diffusion, broadening the initially sharp boundary with time. As a consequence, simulated physical diffusion will be quantitatively in error, and overestimated. After Fig. 3 of [73] by Cox.

bin are taken to arise directly from chemical sedimentation and diffusion fluxes [11, 224, 225].

A key drawback of this approach is that unless Δr is much smaller than the boundary width, or the sedimentation term $\omega^2 s$ is very small, the diffusion is not accurately described. Extra "numerical diffusion" occurs when the translation of the boundary with time is comparable to Δr, caused by averaging (or mixing) of concentrations within the bins, as illustrated in Fig. A.1 Panel B. This effect is well known in other scientific fields where numerical solutions of partial-differential equations of the convection-diffusion type are computed. Errors decrease with increasing N_r, though using a sufficiently high N_r would make this method fairly costly. Different strategies have been developed to improve accuracy, involving the departure from equidistant radial grid. In the approach by Dishon, Weiss, and Yphantis [29], the spacing of grid points is dynamically adjusted to exhibit higher density in regions of higher gradients. Cox developed schemes of non-equidistant and time-dependent radial grid points $r_i(t)$ that eliminate sedimentation flows between bins [73, 227].

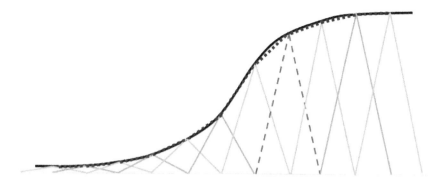

Figure A.2 In the finite difference approach by Claverie [223] the sedimentation boundary (bold black line) is approximated as sums of triangular hat functions $P_i(r)$, a particular one highlighted by blue dashed lines. Multiplied with proper concentration factors, the sum of triangular elements produces a piece-wise linear approximation of the boundary (dotted red line).

A.3 THE FINITE ELEMENT APPROACH

In 1975 Claverie [223] introduced a completely different approach that is not based on physically inspired well-mixed bins and therefore avoids numerical diffusion. Instead it relies on a mathematical approximation of concentration profiles by piecewise linear segments. This is achieved as the sum of triangular elements $P_i(r)$

$$
P_i(r) = \begin{cases} \frac{r - r_{i-1}}{r_i - r_{i-1}} & \text{for } r_{i-1} \le r \le r_i \\ \frac{r_{i+1} - r}{r_{i+1} - r_i} & \text{for } r_i < r \le r_{i+1} \\ 0 & \text{else} \end{cases}, \tag{A.1}
$$

for example, the blue dashed one illustrated in Fig. A.2. Solving the Lamm equation corresponds to the determination of suitable amplitudes $\tilde{\chi}_i(t_j)$ for the hat functions such that

$$
\chi(r, t_j) \approx \sum_i \tilde{\chi}_i(t_j) P_i(r) \tag{A.2}
$$

based on criteria outlined above.[2] This is a fairly basic special case of general finite element methods, for which a large literature exists in applied mathematics and engineering. The coefficients can be determined by determining the mass balance of flows in and out of the regions of specific elements P_k through radial integration of the Lamm equation Eq. (2.30)

$$
\int_m^b \frac{\partial \chi}{\partial t} P_k(r) r dr = - \int_m^b \frac{1}{r} \frac{\partial}{\partial r} \left(\chi s \omega^2 r^2 - D \frac{\partial \chi}{\partial r} r \right) P_k(r) r dr . \tag{A.3}
$$

Integration by parts of the r.h.s., taking advantage of the vanishing fluxes in the case of impermeable boundary conditions at the ends of the solution column, leads

---

[2]In the following we will skip over critically important but nitty-gritty details of time steps and spatial end effects for the arrays (such as at $i = 1$ and $i = N_r$).

Figure A.3 Schematics of a moving grid (vertical bars) at different time points (rows). Except for the first and last point, which are at the meniscus and bottom, respectively, the grid is spaced as prescribed by the power-law Eq. (A.9). At later times all points migrate like a sedimenting particle with s-value $s_g = s$, which renders sedimentation fluxes zero. After the time Δt^* the movement maps each point to the location of the initial right neighbor (vertical thin lines as a visual guide). The point $\#N_r - 1$ is removed (indicated by a half-sized bar) and a new point is inserted in the beginning (indicated by a half-sized bar), such that a grid identical to the starting grid is obtained after renumbering the indices.

to

$$\int_m^b \frac{\partial \chi}{\partial t} P_k(r) r dr = s\omega^2 \int_m^b \chi \frac{\partial P_k}{\partial r} r^2 dr - D \int_m^b \frac{\partial P_k}{\partial r} \frac{\partial \chi}{\partial r} r dr . \tag{A.4}$$

Inserting the approximation Eq. (A.2) we arrive at an equation system for the coefficients

$$0 = \sum_i \frac{\partial \tilde{\chi}_i}{\partial t} \int_m^b P_i P_k r dr - s\omega^2 \sum_i \tilde{\chi}_i \int_m^b P_i \frac{\partial P_k}{\partial r} r^2 dr + D \sum_i \tilde{\chi}_i \int_m^b \frac{\partial P_i}{\partial r} \frac{\partial P_k}{\partial r} r dr , \tag{A.5}$$

where the integrals over hat functions and their derivatives can be carried out analytically ahead of time and results arranged in matrices $\mathbf{B}$, $\mathbf{A}^{(2)}$, and $\mathbf{A}^{(1)}$ (terms left to right). Thus, we can write the propagation as a matrix equation

$$0 = \sum_i \frac{\partial \tilde{\chi}_i}{\partial t} \mathbf{B}_{ik} - s\omega^2 \sum_i \tilde{\chi}_i \mathbf{A}_{ik}^{(2)} + D \sum_i \tilde{\chi}_i \mathbf{A}_{ik}^{(1)} \tag{A.6}$$

that can be easily solved, with suitable time steps and temporal predictor-corrector schemes providing stability and error control. Time-dependent rotor speeds $\omega(t)$ can be accommodated naturally through updating of the propagation matrices. In fact, inclusion of the initial acceleration of the rotor into the model is numerically advantageous, as it limits steep gradients at small D, and thereby improves numerical stability in this case.

A.4   FINITE ELEMENTS ON A MOVING GRID

The original finite element method by Claverie [223] was developed with a stationary and equidistant grid in mind, but it can be similarly applied for other grids.

In particular, a moving grid was introduced [74] where the radial grid points are functions of time $r_i(t)$ in a way that obeys the law of sedimentation of a particle

$$r_i(t) = r_{i,0} e^{s_g \omega^2 (t - t_0)} \qquad (A.7)$$

with a "grid sedimentation coefficient" s_g. This causes a time dependence of the hat functions $P_i(r, t)$ generating an additional term for Eq. (A.5) with coefficients that are equally easy to integrate analytically [74]. With the particular choice of grid movement, on the frame of reference of the grid, no sedimentation occurs with the choice $s_g = s$, thus simplifying the computational effort to migration from diffusion.[3]

Further, the initial grid can be chosen non-equidistant as

$$r_{i,0}(t) = m \left(\frac{b}{m} \right)^{\frac{i-3/2}{N_r-1}}, \qquad (A.8)$$

which has the interesting property that after a certain time interval of propagation

$$\Delta t^* = \frac{\ln(b/m)}{\omega^2 s_g (N_r - 1)} \qquad (A.9)$$

it occurs that

$$r_i(t + \Delta t^*) = r_{i+1}(t), \qquad (A.10)$$

i.e., the radial grid precisely maps onto itself. The significance of this was shown in [74]: Where a time step Δt^* can be used[4] in conjunction with $s_g = s$, the simulation of sedimentation elegantly reduces to incrementing the vector indices by 1 (Fig. A.3). End effects from the movement of the grid can be addressed efficiently and accurately through mass conservation considerations [74].

This moving grid approach is particularly useful for the simulation of sedimentation of very large particles, which would require very costly fine discretization for the standard Claverie approach to achieve numerically stable results. Under these conditions, permeable boundary conditions are applicable, which remove inefficiencies from modeling back-diffusion and leads to trivial and exact values at the bottom.[5]

---

[3]This aspect is reminiscent of the moving bins of Cox [73], but now carried out in the finite element framework with a grid that is explicitly designed for this purpose. In contrast to the method of Cox, the moving grid Eq. (A.7) will not accommodate concentration-dependent s-values in the sedimenting frame of reference — in this case reduced sedimentation fluxes will occur in different regions of the cell where $s(c) \neq s_g$.

[4]This may not always be the case, due to considerations of error limits, fixed required report times, rotor acceleration schedules imposing periodic updates on the propagation matrices, etc. Generally an algorithm will consider these factors at any given time and make a dynamic selection of Δt.

[5]A method that uses the same moving grid but applies improved space-time discretization in the steep back-diffusion region close to the bottom was later described by Cao and Demeler, referred to as ASTFEM [226]. The emphasis in the bottom region is theoretically interesting but irrelevant in practice because the region is entirely unobservable for large particles, and therefore does neither need to be fit nor simulated.

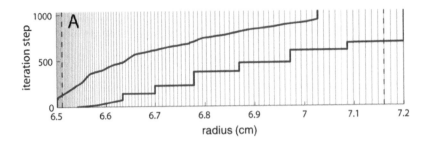

Figure A.4 Schematics of an adaptive grid with error control and active/inactive computation. The grid points (vertical bars) are chosen with square-root increase of Δr_i following Eqs. (A.12) and (A.13), except between the meniscus at 6.5 cm and the left fitting limit at 6.51 cm (left blue dashed line), where the grid points are equidistant. The vertical axis displays the iteration steps necessary for the simulated sedimentation of a 10 S, 450 kDa species. Highlighted in red are the "active" regions of the boundary, where computational steps are required, and black are "inactive" radial points that are within 10^{-4} of the trivial analytical predictions for the solution and solvent plateaus. Reproduced from [34].

> The Lamm equation options box allows the user to apply the moving hat finite element algorithm, and to automatically switch to the non-moving grid below an adjustable s-value threshold.

A.5   ADAPTIVE GRIDS AND ERROR CONTROL

For simulating sedimentation in the regime of diffusion-broadened boundaries with low to moderate s-values, a different grid spacing has been developed [34] that excels at predicting simulation errors as a function of the grid size N_r. This makes it possible to determine the necessary N_r in an adaptive fashion that is dependent on a given error tolerance for the maximum error in the boundary. N_r will adapt depending on the given sedimentation parameters s and D, as well as the end points and fitting limits of the solution column. This unique aspect makes this algorithm considerably more efficient than previous algorithms where N_r was preset empirically, often either too conservatively with wastefully high N_r or — worse — carelessly too coarse with large ensuing discretization errors [38].

A key consideration in this algorithm is the obligate error that occurs when approximating a smooth curved function with piece-wise linear segments, such as the red dotted line in Fig. A.2. Clearly, the more segments can be used the better the approximation can be. Likewise, to achieve the same accuracy, a higher density of grid points is needed for steeper boundaries with higher curvature. Importantly, the boundaries will always exhibit an approximately similar shape of error functions, translated and stretched during sedimentation. As shown in detail in [34] a relative precision of ∼0.001 can be achieved by a piece-wise linear description of these sedimentation boundaries if 25 grid points are placed across the central 90% of the boundary.

The steepest boundary is observed at early times when the boundary is at the smallest radius that is experimentally observed, which for the purpose of the simulation can be taken as the left fitting limit r_1^*. The time required for the boundary to reach this point can be calculated, on the basis of the given sedimentation coefficient, as $\tau \approx (r_1^* - m)/sw^2 r_1^*$. Taking advantage of the $\sqrt{t}$ dependence of diffusion, we can obtain a measure of the boundary spread

$$\sigma_{\text{boundary}} \approx \frac{1}{\omega}\sqrt{\frac{D}{s}} \times \frac{r_1^* - m}{r_1^*} . \tag{A.11}$$

It is useful to scale the density of grid points relative to this quantity σ_{boundary}. With a scaling factor α, defined as $\Delta r = \sigma_{\text{boundary}}/\alpha$, for example, 25 grid points across the central 90% are achieved with a value $\alpha = 5$. Since the boundary spreads with time, we can lower the density of grid points encountered by the boundary at later times, while keeping the obligate discretization errors at the same level. From Eq. (A.11), the required grid point density is a function of radius. Therefore, a grid can be consecutively built up as

$$r_i = \begin{cases} m & \text{for } i = 1 \\ m + \sum_{j<i} \Delta r_j & \text{for } i = 2 \ldots N_r - 1 \\ b & \text{for } i = N_r \end{cases} \tag{A.12}$$

using the spacing

$$\Delta r_i = \frac{1}{\alpha\omega}\sqrt{\frac{D}{s}} \times \begin{cases} \sqrt{\frac{r_1^* - m}{r_1^*}} & \text{for } r \le r_1^* \\ \sqrt{\frac{r_i - m}{r_i}} & \text{else} \end{cases} , \tag{A.13}$$

which adjusts to achieve a constant density of points relative to the boundary spread during the sedimentation process. N_r is determined as a result of the iteration in Eqs. (A.12) and (A.13) as the number where the next step would otherwise exceed the bottom radius. An example of the resulting can be found in Fig. A.4.

By default, the Lamm equation solver in SEDFIT and SEDPHAT will automatically switch between these algorithms to achieve the best possible efficiency while ensuring accuracy by default to within 0.001 or better. In the Lamm equation parameter box, this is indicated by a radio button automatic finite element algorithm. When the adaptive grid is active, this is indicated by a checkmark automatic dynamic grid size. Optionally, the value of α can be entered in the field points per sigma, alongside upper and lower limits for N_r.

It is termed an adaptive grid in a sense that it adjusts to the particular sedimentation parameters and geometry to be simulated, and is calculated prior to the Lamm equation solution. It naturally lends itself to be combined with an estimate of the extent of back-diffusion to determine whether reflective boundary conditions are necessary, as well as the inactivation of iterative computation in plateau regions (Fig. A.4). Due to the square-root spacing of the adaptive grid it cannot usefully sediment like the moving grid, but in the finite element implementation, the adaptive grid shows excellent performance in terms of accuracy and efficiency. Therefore the finite element method with adaptive grid is optimal for cases of low to medium s-values, whereas for high-s, low-D simulations, the moving grid finite element method described above is advantageous.

Calculating Distributions

DISTRIBUTIONS can be calculated after discretization of the distribution problem and the reduction to linear equation systems. More complex questions arise in the context of multi-component distribution analysis with spectral subspaces, regularization, mass conservation, and statistical error analysis. The following sections briefly sketch the computational background to highlight how calculating distribution leads to well-established standard problems of matrix algebra and numerical mathematics.

B.1 BASIC FRAMEWORK

The basic scheme starts at the minimization problem stated already in Section 3.2.3. For clarity we leave out systematic noise contributions, as well as normalization factors; both can be introduced as described in Section 1.1.4.1 without any further complications. We abbreviate data $a(r_i, t_j)$ as a_{ij}, the Lamm equation solution models $\chi_1(s_l, r_i, t_j)$ that represent the kernel as $\chi_{l,ij}$, and the unknown distribution values $\gamma(s_l)$ as γ_l, so that we can rewrite Eq. (3.5) from p. 63 in a more compact form:

$$\underset{\gamma_l \geq 0}{Min} \sum_{i,j} \left[a_{ij} - \sum_{l=1}^{N} \gamma_l \chi_{l,ij} \right]^2 . \tag{B.1}$$

For calculating the unknown γ_l we take the partial derivative with regard to a particular γ_n, which must vanish in the minimum

$$0 = \frac{\partial}{\partial \gamma_n} \sum_{i,j} \left[a_{ij} - \sum_{l=1}^{N} \gamma_l \chi_{l,ij} \right]^2 . \tag{B.2}$$

This holds true for any $n = 1 \ldots N$. With the chain rule we have

$$0 = 2 \sum_{i,j} \left[a_{ij} - \sum_{l=1}^{N} \gamma_l \chi_{l,ij} \right] \left(-\chi_{n,ij} \right), \tag{B.3}$$

which after changing the order of summation leads to

$$\sum_{i,j} a_{ij} \chi_{n,ij} = \sum_{l=1}^{N} \gamma_l \sum_{i,j} \chi_{l,ij} \chi_{n,ij} , \tag{B.4}$$

or, in vector notation,

$$\vec{b} = \mathbf{A}\vec{\gamma}$$

$$b_n = \sum_{i,j} a_{ij} \chi_{n,ij}$$

$$A_{mn} = \sum_{i,j} \chi_{m,ij} \chi_{n,ij} . \tag{B.5}$$

In fact, with these definitions the original minimization problem can be re-stated as

$$\underset{\gamma_l \geq 0}{Min} \left(a_2 - 2\vec{b}\vec{\gamma} + \vec{\gamma}\mathbf{A}\vec{\gamma} \right) \tag{B.6}$$

with a_2 abbreviating the sum of squared data points $\sum_{i,j} a_{ij}^2$ (which is constant and could therefore also be left out in (B.6)).

In principle this can be easily solved with a variety of methods [228]. However, in general it would include negative concentrations γ_l that are physically impossible. To address this problem, there are several algorithms for non-negativity-constrained least squares problems, the most well-known one being NNLS by Lawson and Hanson [229]. It is an "active set method," which reduces the equation system to a subset of unknowns for which only zero or positive values are obtained, by eliminating all rows and columns related to all others with active constraints. The set of indices with active constraints that produces the smallest sum of squares can be found iteratively, based on gradients of the error surface. A strict mathematical proof of convergence to the lowest minimum is found in [229]. The NNLS algorithm involves repeated solution of subsets of the system Eq. (B.5), and can be efficiently implemented [14, 137].

The scale of the problem in relation to the amount of data and size of the model warrants consideration. Most of the computational effort is required for the summations in Eq. (B.5), which scale with N^2 and linearly with the total number of data points. However, these summations can be multi-threaded, taking advantage of multi-processor systems. Computational requirements are discussed in [137] — implementation of large systems is easily possible on personal computers.

Regularization is incorporated in Eq. (B.6) by addition of an entropy term

$$\underset{\gamma_l \geq 0}{Min} \left(a_2 - 2\vec{b}\vec{\gamma} + \vec{\gamma}\mathbf{A}\vec{\gamma} + \alpha \sum \gamma_l \ln \gamma_l \right) \tag{B.7}$$

or a band-matrix $\mathbf{B}$ representing derivatives in Tikhonov–Phillips regularization

$$\vec{b} = (\mathbf{A} + \alpha\mathbf{B}) \vec{\gamma} \tag{B.8}$$

both coupled to F-statistics to control the scaling parameter α. Since the former amounts to a non-linear optimization for each α, whereas the latter remains a linear system, maximum entropy regularization is somewhat slower than Tikhonov–Phillips regularization.

An effective way to estimate errors in the distribution and derived quantities are Monte-Carlo simulations. Since this requires the accurate determination of quantiles of rare events, a large number of iterations are required, typically on the order of 1000. Favorable to this method is the fact that only the inhomogeneity of the system $\vec{b}$ is dependent on the actual data in Eq. (B.5). This reduces the computational effort for Monte-Carlo simulations considerably, since none of the calculations for $\mathbf{A}$ have to be repeated. As a consequence, a large number of iterations can be carried out relatively quickly.

> The **Statistics** menu of **SEDFIT** provides the option to carry out Monte-Carlo simulations in two modes — the determination of error contour lines, and the determination of errors in the integrals of total signal and s_w in different s-ranges.

For error analysis, in practice non-linear parameters often occur besides the distribution itself. This may include, for example, meniscus or frictional ratio parameters. These will affect the Lamm equation solutions $\chi_{k,ij}$, which would cause a dramatic increase of computational load if their variation were to be included in a Monte-Carlo statistical analysis, since computation of a new matrix $\mathbf{A}$ would be required in each step. This may be circumvented by carrying out separate error analysis of non-linear parameters, e.g., using standard strategies of error surface projections and F-statistics [71, 72, 132], to determine confidence limits of the non-linear parameters given experimental data. These errors may be propagated with those from statistical variation of distribution analysis.

B.2   MULTI-COMPONENT ANALYSIS

The basic minimization problem corresponding to the definition of the multi-signal distribution Eq. (7.1) is, after discretization,

$$\underset{c_{kl} \geq 0}{Min} \sum_{\lambda} \sum_{i,j} \left[a_{\lambda,ij} - \sum_{k} \epsilon_{k,\lambda} \sum_{l} c_{k,l} \chi_{l,ij} \right]^2 \tag{B.9}$$

again dropping the optical pathlength, normalization, and noise terms, and for simplicity assuming the Lamm equation solutions $\chi_{l,ij}$ to be identical for the different signals.[1] The data at each signal $a_\lambda(r,t)$ are abbreviated $a_{\lambda,ij}$, and all other symbols are as defined in Section 7.1.

Finding the best-fit concentrations $c_{k,l}$ for the species with sedimentation coefficient s_l for each component k proceeds in the same way as in the one-dimensional case, by requiring the partial derivative of the sum of squares with respect to a particular species n of component κ to vanish:

$$0 = \frac{\partial}{\partial c_{\kappa,n}} \sum_\lambda \sum_{i,j} \left[a_{\lambda,ij} - \sum_k \epsilon_{k,\lambda} \sum_l c_{k,l} \chi_{l,ij} \right]^2 , \qquad (B.10)$$

which leads to

$$0 = 2 \sum_\lambda \sum_{i,j} \left[a_{\lambda,ij} - \sum_k \epsilon_{k,\lambda} \sum_l c_{k,l} \chi_{l,ij} \right] \left(-\epsilon_{\kappa,\lambda} \chi_{n,ij} \right) . \qquad (B.11)$$

Using the abbreviation for $\mathbf{A}$ in Eq. (B.5), and in analogy a vector $\vec{b}_\lambda$ with elements $b_{\lambda,n} = \sum_{i,j} a_{\lambda,ij} \chi_{n,ij}$, , and introducing the spectral matrix $\mathbf{F}$ with

$$F_{k,\kappa} = \sum_\lambda \epsilon_{k,\lambda} \epsilon_{\kappa,\lambda} , \qquad (B.12)$$

we can simplify Eq. (B.11) to

$$\sum_\lambda \epsilon_{\kappa,\lambda} b_{\lambda,n} = \sum_k F_{k,\kappa} \sum_l c_{k,l} A_{l,n} \qquad (B.13)$$

for any combination of $\kappa = 1 \ldots K$ and $n = 1 \ldots N$. This is a multi-dimensional extension of the linear equation system Eq. (B.5) for the unknowns $c_{k,l}$ and can be solved in exactly the same way.

However, the factorization into the sedimentation part $\mathbf{A}$ and the spectral part $\mathbf{F}$ allows us to highlight spectral requirements for components to be distinguishable if they sediment synchronously with the same sedimentation coefficient s_x. If they are the only species, Eq. (B.13) would reduce to

$$\sum_\lambda \epsilon_{\kappa,\lambda} \frac{b_{\lambda,x}}{A_{x,x}} = \sum_k F_{k,\kappa} c_{k,x} \qquad (B.14)$$

with $\mathbf{F}$ being the remaining relevant matrix. The relative error in the l.h.s. will be amplified into relative errors of concentrations by the condition number of $\mathbf{F}$ [228]. For given extinction coefficients, the condition number can be easily calculated, and the l.h.s. can be estimated given common experimental data acquisition errors

---

[1]In practice this is usually not the case, even if the model assigns the same number and species parameters for each component, due to unavoidable imperfections in the data acquisition. This makes the notation and computation more tedious but not conceptually different. Therefore, this complication is omitted here.

and simulated Lamm equation solutions. This leads to the prediction of statistical concentration errors in MSSV Eq. (7.4).

For monochromatic multi-component analysis, the problem is slightly different. The discretized minimization problem corresponding to the definition of the MCMC distribution Eq. (7.8) is lacking the summation over multiple data sets, but gaining an s-value and time dependence of the extinction coefficient, here abbreviated $\epsilon_{k,l,j}$

$$\underset{c_{kl} \geq 0}{Min} \sum_{i,j} \left[a_{ij} - \sum_{k,l} \epsilon_{k,l,j} c_{k,l} \chi_{l,ij} \right]^2 . \tag{B.15}$$

The s-value dependence is described by Eq. (1.5) on p.7, and more extensively in Part I, Section 4.3.2.3. However, the algebra in MCMC is very similar to MSSV, and the key result corresponding to Eq. (B.13) is

$$\sum_{i,j} \epsilon_{\kappa,m,j} a_{ij} \chi_{m,ij} = \sum_{k,l} c_{k,l} \sum_{j} \epsilon_{k,l,j} \epsilon_{\kappa,m,j} \sum_{i} \chi_{l,ij} \chi_{m,ij} \tag{B.16}$$

for any combination of $\kappa = 1 \ldots K$ and $n = 1 \ldots N$. Once again, this is a linear system that can be solved with the usual methods ensuring non-negativity. In analogy, the matrix $\mathbf{F}$ from Eq. (B.12) governing the error propagation is now

$$F_{k,\kappa} = \sum_{j} \epsilon_{k,l,j} \epsilon_{\kappa,m,j} \sum_{i} \chi_{l,ij} \chi_{m,ij} , \tag{B.17}$$

which has a dependence on the s-values. For species of the same s-value s_x to be distinguished, the dominant quantity is the sum over the mutual product of extinction coefficients, as discussed in Section 7.2 Eq. (7.9). Calculations with spatio-temporal signal modulation proceed analogously.

B.3   DISTRIBUTION ANALYSIS WITH MASS CONSERVATION

A slight variation of the basic scheme allows us to incorporate mass conservation constraints into the MSSV calculation (Section 7.1.5). If we know the total mass of material of a component k to be C_k^{tot}, a discretized least-squares minimization corresponding to the coupled equation system Eq. (7.7) is

$$\underset{c_{kl} \geq 0}{Min} \left\{ \sum_{\lambda} \sum_{i,j} \left[a_{\lambda,ij} - \sum_{k} \epsilon_{k,\lambda} \sum_{l} c_{k,l} \chi_{l,ij} \right]^2 + \alpha \sum_{k} \left[C_k^{tot} - \sum_{l} c_{k,l} \right]^2 \right\} . \tag{B.18}$$

For clarity of presentation we neglect potential limited integration ranges in the mass conservation considerations, assume we have constraints for all components k, and avoid additional factors that can control the relative weight of the constraints [37]. Here the mass conservation criterion is embedded as a second term, balanced like regularization factors with a multiplicative α, which adjusts how stringent mass conservation must be fulfilled relative to the goal of achieving the best possible fit. Typically this judgment can be made with the help of F-statistics, and/or by considering experimental uncertainties in the C_k^{tot}.

For any particular value of α, the derivative with regard to a component concentration $c_{\kappa,n}$ gains a term compared to Eq. (B.11)

$$
\begin{aligned}
0 =& 2 \sum_\lambda \sum_{i,j} \left[a_{\lambda,ij} - \sum_k \epsilon_{k,\lambda} \sum_l c_{k,l} \chi_{l,ij} \right] (-\epsilon_{\kappa,\lambda} \chi_{n,ij}) \\
& - 2\alpha \sum_k \left[C_k^{tot} - \sum_l c_{k,l} \right],
\end{aligned}
\tag{B.19}
$$

which leads to the following extension of Eq. (B.13)

$$
\sum_\lambda \epsilon_{\kappa,\lambda} b_{\lambda,n} + \alpha \sum_k C_k^{tot} = \sum_{k,l} \left(F_{k,\kappa} + \frac{\alpha}{A_{l,n}} \right) c_{k,l} A_{l,n}
\tag{B.20}
$$

for $\kappa = 1 \ldots K$ and $n = 1 \ldots N$. Thus, responsible for the discrimination of species with the same s-value s_x is no longer only the spectral matrix $\mathbf{F}$, but now the sum $\mathbf{F} + \alpha/A_{x,x}\mathbf{I}$. The new diagonal component will cause the condition of the matrix to improve. In this sense the new mass conservation constraints will fulfill the same purpose as spectral discrimination. However, if α becomes very large, the new term will decrease the resolution of different s-values, as in the limit $\alpha \to \infty$ the first term on the l.h.s. and the first terms in the parenthesis of the r.h.s. disappear and the $A_{l,n}$ cancel out. On the other hand, a better situation arises when mass conservation constraints are more specifically assigned to limited regions of s-values, which will be reflected in a different structure and condition of the equation system Eq. (B.20).

Bibliography

[1] P. Schuck, H. Zhao, C.A. Brautigam, and R. Ghirlando, *Basic Principles of Analytical Ultracentrifugation.* Boca Raton, FL: CRC Press, 2015. ISBN 978-1-49-875115-5

[2] H.K. Schachman, *Ultracentrifugation in Biochemistry.* New York: Academic Press, 1959. ISBN 1483270947

[3] T. Svedberg and K.O. Pedersen, *The Ultracentrifuge.* London: Oxford University Press, 1940. ISBN 0384588905

[4] R. Trautman, S.P. Spragg, and H.H. Halsall, "Absorption optics data processing with standards errors for sedimentation and diffusion coefficients from moving boundary ultracentrifugation." *Anal. Biochem.*, vol. 28, no. 1, pp. 396–415, 1969. doi: 10.1016/0003-2697(69)90195-X

[5] G.A. Gilbert and L.M. Gilbert, "Ultracentrifuge studies of interactions and equilibria: Impact of interactive computer modelling." *Biochem. Soc. Trans.*, vol. 8, no. 5, pp. 520–522, 1980. doi: 10.1042/bst0080520

[6] R. Cohen and J.M. Claverie, "Sedimentation of generalized systems of interacting particles. II. Active enzyme centrifugation–theory and extensions of its validity range," *Biopolymers*, vol. 14, no. 8, pp. 1701–1716, 1975. doi: 10.1002/bip.1975.360140812

[7] D.J. Cox, "Computer simulation of sedimentation in the ultracentrifuge I. Diffusion," *Arch. Biochem. Biophys.*, vol. 112, no. 2, pp. 249–258, 1965. doi: 10.1016/0003-9861(65)90043-3

[8] G.P. Todd and R.H. Haschemeyer, "General solution to the inverse problem of the differential equation of the ultracentrifuge," *Proc. Natl. Acad. Sci. USA*, vol. 78, no. 11, pp. 6739–6743, 1981.

[9] L.A. Holladay, "Simultaneous rapid estimation of sedimentation coefficient and molecular weight," *Biophys. Chem.*, vol. 11, no. 2, pp. 303–308, 1980. doi: 10.1016/0301-4622(80)80033-0

[10] J.S. Philo, "An improved function for fitting sedimentation velocity data for low-molecular-weight solutes," *Biophys. J.*, vol. 72, no. 1, pp. 435–444, 1996. doi: 10.1016/S0006-3495(97)78684-3

[11] P. Schuck, C.E. MacPhee, and G.J. Howlett, "Determination of sedimentation coefficients for small peptides," *Biophys. J.*, vol. 74, no. 1, pp. 466–474, 1998. doi: 10.1016/S0006-3495(98)77804-X

[12] B. Demeler, J. Behlke, and O. Ristau, "Molecular parameters from sedimentation velocity experiments: Whole boundary fitting using approximate and numerical solutions of Lamm equation," *Methods Enzymol.*, vol. 321, no. 1998, pp. 38–66, 2000. doi: 10.1016/S0076-6879(00)21186-5

[13] P. Schuck, "Analytical ultracentrifugation as a tool for studying protein interactions," *Biophys. Rev.*, vol. 5, no. 2, pp. 159–171, 2013. doi: 10.1007/s12551-013-0106-2

[14] P. Schuck, "Size-distribution analysis of macromolecules by sedimentation velocity ultracentrifugation and Lamm equation modeling," *Biophys. J.*, vol. 78, no. 3, pp. 1606–1619, 2000. doi: 10.1016/S0006-3495(00)76713-0

[15] P. Schuck and B. Demeler, "Direct sedimentation analysis of interference optical data in analytical ultracentrifugation." *Biophys. J.*, vol. 76, no. 4, pp. 2288–2296, 1999. doi: 10.1016/S0006-3495(99)77384-4

[16] H. Zhao and P. Schuck, "Global multi-method analysis of affinities and cooperativity in complex systems of macromolecular interactions." *Anal. Chem.*, vol. 84, no. 21, pp. 9513–9519, 2012. doi: 10.1021/ac302357w

[17] D.J. Cox, "Sedimentation of an initially skewed boundary," *Science*, vol. 152, no. 3720, pp. 359–361, 1966. doi: 10.1126/science.152.3720.359

[18] P. Schuck, "A model for sedimentation in inhomogeneous media. I. Dynamic density gradients from sedimenting co-solutes," *Biophys. Chem.*, vol. 108, no. 1-3, pp. 187–200, 2004. doi: 10.1016/j.bpc.2003.10.016

[19] P. Schuck, "A model for sedimentation in inhomogeneous media. II. Compressibility of aqueous and organic solvents," *Biophys. Chem.*, vol. 108, no. 1-3, pp. 201–214, 2004. doi: 10.1016/j.bpc.2003.10.017

[20] P. Schuck, Z. Taraporewala, P. McPhie, and J.T. Patton, "Rotavirus nonstructural protein NSP2 self-assembles into octamers that undergo ligand-induced conformational changes." *J. Biol. Chem.*, vol. 276, no. 13, pp. 9679–9687, 2001. doi: 10.1074/jbc.M009398200

[21] J. Ma, M. Metrick, R. Ghirlando, H. Zhao, and P. Schuck, "Variable-field analytical ultracentrifugation: I. Time-optimized sedimentation equilibrium," *Biophys J.*, vol. 109, no. 4, pp. 827–837, 2015. doi: 10.1016/j.bpj.2015.07.015

[22] J. Ma, H. Zhao, J. Sandmaier, J.A. Liddle, and P. Schuck, "Variable-field analytical ultracentrifugation: II. Gravitational sweep sedimentation," *Biophys. J.*, vol. 110, no. 1, pp. 103–112, 2016. doi: 10.1016/j.bpj.2015.11.027

[23] H. Faxén, "Über eine Differentialgleichung aus der physikalischen Chemie," *Ark. Mat. Astr. Fys.*, vol. 21B, pp. 1–6, 1929.

[24] H. Fujita, *Foundations of Ultracentrifugal Analysis*. New York: John Wiley & Sons, 1975.

[25] G.H. Weiss and D.A. Yphantis, "Rectangular approximation for concentration-dependent sedimentation in the ultracentrifuge," *J. Chem. Phys.*, vol. 42, no. 6, pp. 2117–2123, 1965. doi: 10.1063/1.1696254

[26] L.A. Holladay, "Molecular weights from approach-to-sedimentation equilibrium data using nonlinear regression analysis," *Biophys. Chem.*, vol. 10, pp. 183–185, 1979.

[27] J. Behlke and O. Ristau, "A new approximate whole boundary solution of the Lamm differential equation for the analysis of sedimentation velocity experiments," *Biophys. Chem.*, vol. 95, no. 1, pp. 59–68, 2002. doi: 10.1016/S0301-4622(01)00248-4

[28] D.A. Yphantis and D.F. Waugh, "Transient solute distributions from the basic equation of the ultracentrifuge," *J. Phys. Chem.*, vol. 57, no. 3, pp. 312–318, 1953. doi: 10.1021/j150504a012

[29] M. Dishon, G.H. Weiss, and D.A. Yphantis, "Numerical solutions of the Lamm equation. I. Numerical procedure," *Biopolymers*, vol. 4, no. 4, pp. 449–455, 1966. doi: 10.1002/bip.1966.360040406

[30] M. Dishon, G.H. Weiss, and D.A. Yphantis, "Numerical solutions of the Lamm equation. III. Velocity centrifugation," *Biopolymers*, vol. 5, no. 8, pp. 697–713, 1967. doi: 10.1002/bip.1967.360050804

[31] M. Dishon, G.H. Weiss, and D.A. Yphantis, "Numerical solutions of the Lamm equation. V. Band centrifugation," *Ann. N.Y. Acad. Sci.*, vol. 164, no. 2, pp. 33–51, 1969. doi: 10.1111/j.1749-6632.1969.tb14031.x

[32] M. Dishon, G.H. Weiss, and D.A. Yphantis, "Numerical solutions of the Lamm equation. VI. Effects of hydrostatic pressure on velocity sedimentation of two-component systems." *J. Polym. Sci.*, vol. 8, no. 12, pp. 2163–2175, 1970. doi: 10.1002/pol.1970.160081212

[33] M. Dishon, G.H. Weiss, and D.A. Yphantis, "Kinetics of sedimentation in a density gradient," *Biopolymers*, vol. 10, no. 1, pp. 2095–2111, 1971. doi: 10.1002/bip.360101107

[34] P.H. Brown and P. Schuck, "A new adaptive grid-size algorithm for the simulation of sedimentation velocity profiles in analytical ultracentrifugation," *Comput. Phys. Commun.*, vol. 178, no. 2, pp. 105–120, 2008. doi: 10.1016/j.cpc.2007.08.012

[35] A. Balbo, K.H. Minor, C.A. Velikovsky, R.A. Mariuzza, C.B. Peterson, and P. Schuck, "Studying multi-protein complexes by multi-signal sedimentation velocity analytical ultracentrifugation," *Proc. Natl. Acad. Sci. USA*, vol. 102, no. 1, pp. 81–86, 2005. doi: 10.1073/pnas.0408399102

[36] J. Walter, K. Löhr, E. Karabudak, W. Reis, J. Mikhael, W. Peukert, W. Wohlleben, and H. Cölfen, "Multidimensional analysis of nanoparticles with highly disperse properties using multiwavelength analytical ultracentrifugation." *ACS Nano*, vol. 8, no. 9, pp. 8871–8886, 2014. doi: 10.1021/nn503205k

[37] C.A. Brautigam, S.B. Padrick, and P. Schuck, "Multi-signal sedimentation velocity analysis with mass conservation for determining the stoichiometry of protein complexes," *PLoS One*, vol. 8, no. 5, p. e62694, 2013. doi: 10.1371/journal.pone.0062694

[38] P.H. Brown, A. Balbo, and P. Schuck, "On the analysis of sedimentation velocity in the study of protein complexes," *Eur. Biophys. J.*, vol. 38, no. 8, pp. 1079–1099, 2009. doi: 10.1007/s00249-009-0514-1

[39] H. Zhao, E. Casillas, H. Shroff, G.H. Patterson, and P. Schuck, "Tools for the quantitative analysis of sedimentation boundaries detected by fluorescence optical analytical ultracentrifugation," *PLoS One*, vol. 8, no. 10, p. e77245, 2013. doi: 10.1371/journal.pone.0077245

[40] H. Zhao, J. Ma, M. Ingaramo, E. Andrade, J. MacDonald, G. Ramsay, G. Piszczek, G.H. Patterson, and P. Schuck, "Accounting for photophysical processes and specific signal intensity changes in fluorescence-detected sedimentation velocity." *Anal. Chem.*, vol. 86, no. 18, pp. 9286–9292, 2014. doi: 10.1021/ac502478a

[41] H. Zhao, Y. Fu, C. Glasser, E. Andrade, M.L. Mayer, G. Patterson, and P. Schuck, "Monochromatic multicomponent fluorescence sedimentation velocity for the study of high-affinity protein interactions, 2016 *eLife*. doi: 10.7554/eLife.17812

[42] D. Prosperi, C. Morasso, F. Mantegazza, M. Buscaglia, L. Hough, and T. Bellini, "Phantom nanoparticles as probes of biomolecular interactions." *Small*, vol. 2, no. 8-9, pp. 1060–1067, 2006. doi: 10.1002/smll.200600106

[43] D.F. Lyons, J.W. Lary, B. Husain, J.J. Correia, and J.L. Cole, "Are fluorescence-detected sedimentation velocity data reliable?" *Anal. Biochem.*, vol. 437, no. 2, pp. 133–137, 2013. doi: 10.1016/j.ab.2013.02.019

[44] E. Betzig, G.H. Patterson, R. Sougrat, O.W. Lindwasser, S. Olenych, J.S. Bonifacino, M.W. Davidson, J. Lippincott-Schwartz, and H.F. Hess, "Imaging intracellular fluorescent proteins at nanometer resolution." *Science*, vol. 313, no. 5793, pp. 1642–1645, 2006. doi: 10.1126/science.1127344

[45] G.H. Patterson, M. Davidson, S. Manley, and J. Lippincott-Schwartz, "Superresolution imaging using single-molecule localization," *Annu. Rev. Phys. Chem.*, vol. 61, no. 1, pp. 345–367, 2010. doi: 10.1146/annurev.physchem.012809.103444

[46] P. Schuck, "Some statistical properties of differencing schemes for baseline correction of sedimentation velocity data," *Anal. Biochem.*, vol. 401, no. 2, pp. 280–287, 2010. doi: 10.1016/j.ab.2010.02.037

[47] H. Zhao, P.H. Brown, A. Balbo, M.C. Fernandez Alonso, N. Polishchuck, C. Chaudhry, M.L. Mayer, R. Ghirlando, and P. Schuck, "Accounting for solvent signal offsets in the analysis of interferometric sedimentation velocity data," *Macromol. Biosci.*, vol. 10, no. 7, pp. 736–745, 2010. doi: 10.1002/mabi.200900456

[48] I.K. MacGregor, A.L. Anderson, and T.M. Laue, "Fluorescence detection for the XLI analytical ultracentrifuge," *Biophys. Chem.*, vol. 108, no. 1-3, pp. 165–185, 2004. doi: 10.1016/j.bpc.2003.10.018

[49] R.R. Kroe and T.M. Laue, "NUTS and BOLTS: Applications of fluorescence-detected sedimentation," *Anal. Biochem.*, vol. 390, no. 1, pp. 1–13, 2009. doi: 10.1016/j.ab.2008.11.033

[50] H. Zhao, R. Ghirlando, G. Piszczek, U. Curth, C.A. Brautigam, and P. Schuck, "Recorded scan times can limit the accuracy of sedimentation coefficients in analytical ultracentrifugation," *Anal. Biochem.*, vol. 437, no. 1, pp. 104–108, 2013. doi: 10.1016/j.ab.2013.02.011

[51] R. Ghirlando, A. Balbo, G. Piszczek, P.H. Brown, M.S. Lewis, C.A. Brautigam, P. Schuck, and H. Zhao, "Improving the thermal, radial, and temporal accuracy of the analytical ultracentrifuge through external references," *Anal. Biochem.*, vol. 440, no. 1, pp. 81–95, 2013. doi: 10.1016/j.ab.2013.05.011

[52] H. Zhao, R. Ghirlando, C. Alfonso, F. Arisaka, I. Attali, D.L. Bain, M.M. Bakhtina, D.F. Becker, G.J. Bedwell, A. Bekdemir, T.M.D. Besong, C. Birck, C.A. Brautigam, W. Brennerman, O. Byron, A. Bzowska, J.B. Chaires, C.T. Chaton, H. Cölfen, K.D. Connaghan, K.A. Crowley, U. Curth, T. Daviter, W.L. Dean, A.I. Diez, C. Ebel, D.M. Eckert, L.E. Eisele, E. Eisenstein, P. England, C. Escalante, J.A. Fagan, R. Fairman, R.M. Finn, W. Fischle, J. García de la Torre, J. Gor, H. Gustafsson, D. Hall, S.E. Harding, J.G. Hernandez Cifre, A.B. Herr, E.E. Howell, R.S. Isaac, S.-C. Jao, D. Jose, S.-J. Kim, B. Kokona, J.A. Kornblatt, D. Kosek, E. Krayukhina, D. Krzizike, E.A. Kusznir, H. Kwon, A. Larson, T.M. Laue, A. Le Roy, A.P. Leech, H. Lilie, K. Luger, J.R. Luque-Ortega, J. Ma, C.A. May, E.L. Maynard, A. Modrak-Wojcik, Y.-F. Mok, N. Mücke, L. Nagel-Steger, G.J. Narlikar, M. Noda, A. Nourse, T. Obsil, C.K. Park, J.-K. Park, P.D. Pawalek, E.E. Perdue, S.J. Perkins, M.A. Perugi, "A multilaboratory comparison of calibration accuracy and the performance of external references in analytical ultracentrifugation," *PLoS One*, vol. 10, no. 5, p. e0126420, 2015. doi: 10.1371/journal.pone.0126420

[53] S.R. Kar, J.S. Kingsbury, M.S. Lewis, T.M. Laue, and P. Schuck, "Analysis of transport experiment using pseudo-absorbance data," *Anal. Biochem.*, vol. 285, no. 1, pp. 135–142, 2000. doi: 10.1006/abio.2000.4748

[54] M.R. Osborne, "Some special nonlinear least squares problems," *Siam J. Numer. Anal.*, vol. 12, no. 4, pp. 571–592, 1975. doi: 10.1137/0712044

[55] H. Boukari, R.J. Nossal, D.L. Sackett, and P. Schuck, "Hydrodynamics of nanoscopic tubulin rings in dilute solution," *Phys. Rev. Lett.*, vol. 93, p. 98106, 2004. doi: 10.1103/PhysRevLett.93.098106

[56] S.W. Provencher, "A constrained regularization method for inverting data represented by linear algebraic or integral equations," *Comput. Phys. Commun.*, vol. 27, no. 3, pp. 213–227, 1982. doi: 10.1016/0010-4655(82)90173-4

[57] P. Schuck, M.A. Perugini, N.R. Gonzales, G.J. Howlett, and D. Schubert, "Size-distribution analysis of proteins by analytical ultracentrifugation: strategies and application to model systems," *Biophys. J.*, vol. 82, no. 2, pp. 1096–1111, 2002. doi: 10.1016/S0006-3495(02)75469-6

[58] P.H. Brown, A. Balbo, and P. Schuck, "Using prior knowledge in the determination of macromolecular size-distributions by analytical ultracentrifugation." *Biomacromolecules*, vol. 8, no. 6, pp. 2011–2024, 2007. doi: 10.1021/bm070193j

[59] P.H. Brown, A. Balbo, and P. Schuck, "A Bayesian approach for quantifying trace amounts of antibody aggregates by sedimentation velocity analytical ultracentrifugation," *AAPS J.*, vol. 10, no. 3, pp. 481–493, 2008. doi: 10.1208/s12248-008-9058-z

[60] S.A. Ali, N. Iwabuchi, T. Matsui, K. Hirota, S. Kidokoro, M. Arai, K. Kuwajima, P. Schuck, and F. Arisaka, "Reversible and fast association equilibria of a molecular chaperone, gp57A, of bacteriophage T4," *Biophys. J.*, vol. 85, no. 4, pp. 2606–2618, 2003. doi: 10.1016/S0006-3495(03)74683-9

[61] J. Dam, C.A. Velikovsky, R.A. Mariuzza, C. Urbanke, and P. Schuck, "Sedimentation velocity analysis of heterogeneous protein-protein interactions: Lamm equation modeling and sedimentation coefficient distributions c(s)," *Biophys. J.*, vol. 89, no. 1, pp. 619–634, 2005. doi: 10.1529/biophysj.105.059568

[62] W.F. Stafford and P.J. Sherwood, "Analysis of heterologous interacting systems by sedimentation velocity: Curve fitting algorithms for estimation of sedimentation coefficients, equilibrium and kinetic constants." *Biophys. Chem.*, vol. 108, no. 1-3, pp. 231–243, 2004. doi: 10.1016/j.bpc.2003.10.028

[63] H. Schmeisser, I. Gorshkova, P.H. Brown, P. Kontsek, P. Schuck, and K. Zoon, "Two interferons alpha influence each other during their interaction with the extracellular domain of human type I interferon receptor subunit 2," *Biochemistry*, vol. 46, no. 50, pp. 14 638–14 649, 2007. doi: 10.1021/bi7012036

[64] C.A. Brautigam, "Using Lamm-Equation modeling of sedimentation velocity data to determine the kinetic and thermodynamic properties of macromolecular interactions." *Methods*, vol. 54, no. 1, pp. 4–15, 2011. doi: 10.1016/j.ymeth.2010.12.029

[65] P.H. Brown, A. Balbo, H. Zhao, C. Ebel, and P. Schuck, "Density contrast sedimentation velocity for the determination of protein partial-specific volumes." *PLoS One*, vol. 6, no. 10, p. e26221, 2011. doi: 10.1371/journal.pone.0026221

[66] H. Zhao and P. Schuck, "Combining biophysical methods for the analysis of protein complex stoichiometry and affinity in SEDPHAT," *Acta Crystallogr D Biol Crystallogr*, vol. D71, pp. 3–14, 2015. doi: 10.1107/S1399004714010372

[67] J. Dam and P. Schuck, "Calculating sedimentation coefficient distributions by direct modeling of sedimentation velocity concentration profiles," *Methods Enzymol.*, vol. 384, no. 301, pp. 185–212, 2004. doi: 10.1016/S0076-6879(04)84012-6

[68] J. Ma, H. Zhao, and P. Schuck, "A histogram approach to the quality of fit in sedimentation velocity analyses," *Anal. Biochem.*, vol. 483, pp. 1–3, 2015. doi: 10.1016/j.ab.2015.04.029

[69] M. Straume and M.L. Johnson, "Analysis of residuals: Criteria for determining goodness-of-fit." *Methods Enzymol.*, vol. 210, pp. 87–105, 1992. doi: 10.1016/0076-6879(92)10007-Z

[70] H. Zhao, A. Balbo, P.H. Brown, and P. Schuck, "The boundary structure in the analysis of reversibly interacting systems by sedimentation velocity." *Methods*, vol. 54, no. 1, pp. 16–30, 2011. doi: 10.1016/j.ymeth.2011.01.010

[71] M.L. Johnson, "Why, when, and how biochemists should use least squares," *Anal. Biochem.*, vol. 225, pp. 215–225, 1992. doi: 10.1016/0003-2697(92)90356-C

[72] M.L. Johnson and M. Straume, "Comments on the analysis of sedimentation equilibrium experiments," in *Modern Analytical Ultracentrifugation: Techniques and Methods*, T.M. Schuster and T.M. Laue, Eds. Boston: Birkhäuser, 1994, pp. 37–65.

[73] D.J. Cox, "Calculation of simulated sedimentation velocity profiles for self-associating solutes," *Methods Enzymol.*, vol. 2, pp. 212–242, 1978. doi: 10.1016/S0076-6879(78)48012-7

[74] P. Schuck, "Sedimentation analysis of noninteracting and self-associating solutes using numerical solutions to the Lamm equation," *Biophys. J.*, vol. 75, no. 3, pp. 1503–1512, 1998. doi: 10.1016/S0006-3495(98)74069-X

[75] P. Schuck, "Sedimentation patterns of rapidly reversible protein interactions." *Biophys. J.*, vol. 98, no. 9, pp. 2005–2013, 2010. doi: 10.1016/j.bpj.2009.12.4336

[76] P. Schuck, "Diffusion of the reaction boundary of rapidly interacting macromolecules in sedimentation velocity." *Biophys. J.*, vol. 98, no. 11, pp. 2741–2751, 2010. doi: 10.1016/j.bpj.2010.03.004

[77] G.M. Pavlov, I. Perevyazko, and U.S. Schubert, "Velocity sedimentation and intrinsic viscosity analysis of polystyrene standards with a wide range of molar masses," *Macromol. Chem. Phys.*, vol. 211, no. 12, pp. 1298–1310, 2010. doi: 10.1002/macp.200900602

[78] G.M. Pavlov, I. Perevyazko, O.V. Okatova, and U.S. Schubert, "Conformation parameters of linear macromolecules from velocity sedimentation and other hydrodynamic methods," *Methods*, vol. 54, no. 1, pp. 124–135, 2011. doi: 10.1016/j.ymeth.2011.02.005

[79] T. Svedberg, "Molecular weight analysis in centrifugal fields," *Science*, vol. 79, no. 2050, pp. 327–332, 1934. doi: 10.1126/science.79.2050.327

[80] R.F. Hoskins, *Delta Functions. Introduction to Generalized Functions*, 2nd ed. Coll House, U.K.: Horwood Publishing, 1999. ISBN 978-1-904275-39-8

[81] T. Svedberg and J.B. Nichols, "Determination of size and distribution of size of particle by centrifugal methods," *J. Am. Chem. Soc.*, vol. 45, pp. 2910–2917, 1923. doi: 10.1021/ja01665a016

[82] W.F. Harrington and G. Kegeles, "Pressure effects in ultracentrifugation of interacting systems," *Methods Enzymol.*, vol. 27, pp. 306–345, 1973. doi: 10.1016/S0076-6879(73)27016-7

[83] H. Fujita, "Effects of hydrostatic pressure upon sedimentation in the ultracentrifuge," *J. Am. Chem. Soc.*, vol. 78, no. 15, pp. 3598–3604, 1956. doi: 10.1021/ja01596a012

[84] P.W. Bridgman, "The viscosity of liquids under pressure," *Proc. Natl. Acad. Sci. USA*, vol. 11, no. 10, pp. 603–606, 1925.

[85] O. Lamm, "Die Differentialgleichung der Ultrazentrifugierung," *Ark. Mat. Astr. Fys.*, vol. 21B(2), pp. 1–4, 1929.

[86] H. Fujita, *Mathematical Theory of Sedimentation Analysis.* New York: Academic Press, 1962.

[87] W.J. Archibald, "An approximate solution of the differential equation of the ultracentrifuge," *J. Appl. Phys.*, vol. 18, no. 4, p. 362, 1947. doi: 10.1063/1.1697659

[88] H. Fujita and V.J. MacCosham, "Extension of sedimentation velocity theory to molecules of intermediate sizes," *J. Chem. Phys.*, vol. 30, pp. 291–298, 1959. doi: 10.1063/1.1729890

[89] L.A. Holladay, "An approximate solution to the Lamm equation," *Biophys. Chem.*, vol. 10, no. 2, pp. 187–190, 1979. doi: 10.1016/0301-4622(79)85039-5

[90] W.J. Archibald, "A demonstration of some new methods of determining molecular weights from the data of the ultracentrifuge," *J. Phys. Chem.*, vol. 51, no. 5, pp. 1204–1214, 1947. doi: 10.1021/j150455a014

[91] P.-Y. Cheng, "On the meniscus image in the ultracentrifuge," *J. Phys. Chem.*, vol. 61, no. 5, pp. 695–696, 1957. doi: 10.1021/j150551a043

[92] P. Schuck and D.B. Millar, "Rapid determination of molar mass in modified Archibald experiments using direct fitting of the Lamm equation," *Anal. Biochem.*, vol. 259, no. 1, pp. 48–53, 1998. doi: 10.1006/abio.1998.2638

[93] M.T. Gehatia, "AFML-TR-64-377: Ultracentrifugation at variable angular velocity - derivation of basic equations," Wright-Patterson Air Force Base Ohio, Tech. Rep., 1965. Online at: http://www.dtic.mil/dtic/tr/fulltext/u2/615966.pdf

[94] R.J. Nossal and G.H. Weiss, "Sedimentation in a time-varying ultracentrifuge," *Anal. Biochem.*, vol. 38, no. 1, pp. 115–120, 1970. doi: 10.1016/0003-2697(70)90161-2

[95] H. Zhao, A. Balbo, H. Metger, R. Clary, R. Ghirlando, and P. Schuck, "Improved measurement of the rotor temperature in analytical ultracentrifugation," *Anal. Biochem.*, vol. 451, pp. 69–75, 2014. doi: 10.1016/j.ab.2014.02.006

[96] W. Scholtan and H. Lange, "Bestimmung der Teilchengrößenverteilung von Latices mit der Ultrazentrifuge," *Kolloid-Z. u. Z. Polym.*, vol. 250, no. 8, pp. 782–796, 1972. doi: 10.1007/BF01498571

[97] W. Mächtle, "High-resolution, submicron particle size distribution analysis using gravitational-sweep sedimentation." *Biophys. J.*, vol. 76, no. 2, pp. 1080–1091, 1999. doi: 10.1016/S0006-3495(99)77273-5

[98] M.S. Runge, T.M. Laue, D.A. Yphantis, M.R. Lifsics, A. Saito, M. Altin, K. Reinke, and R.C. Williams, "ATP-induced formation of an associated complex between microtubules and neurofilaments." *Proc. Natl. Acad. Sci. U. S. A.*, vol. 78, no. 3, pp. 1431–1435, 1981. doi: 10.1073/pnas.78.3.1431

[99] W.F. Stafford and E.H. Braswell, "Sedimentation velocity, multi-speed method for analyzing polydisperse solutions." *Biophys. Chem.*, vol. 108, no. 1–3, pp. 273–9, 2004. doi: 10.1016/j.bpc.2003.10.027

[100] H. Mach and T. Arvinte, "Addressing new analytical challenges in protein formulation development," *Eur. J. Pharm. Biopharm.*, vol. 78, no. 2, pp. 196–207, 2011. doi: 10.1016/j.ejpb.2011.03.001

[101] M. Meselson, F.W. Stahl, and J. Vinograd, "Equilibrium sedimentation of macromolecules in density gradients," *Proc. Natl. Acad. Sci. USA*, vol. 43, no. 7, pp. 581–588, 1957.

[102] M. Meselson and F.W. Stahl, "The replication of DNA in *Escherichia coli*," *Proc. Natl. Acad. Sci. USA*, vol. 44, no. 7, pp. 671–682, 1958. doi: 10.1073/pnas.44.7.671

[103] A. Tiselius, K.O. Pedersen, and T. Svedberg, "Analytical measurements of ultracentrifugal sedimentation," *Nature*, vol. 140, no. 3550, pp. 848–849, 1937. doi: 10.1038/140848a0

[104] H. Gutfreund and A.G. Ogston, "A method for determining the sedimentation constant of material of low molecular weight: Studies on oxidation products of insulin." *Biochem. J.*, vol. 44, no. 2, pp. 163–166, 1949. doi: 10.1042/bj0440163

[105] R.J. Goldberg, "Sedimentation in the ultracentrifuge," *J. Phys. Chem.*, vol. 57, no. 2, pp. 194–202, 1953. doi: 10.1021/j150503a014

[106] R.L. Baldwin, "Sedimentation coefficients of small molecules: Methods of measurement based on the refractive-index gradient curve; the sedimentation coefficient of polyglucose A." *Biochem. J.*, vol. 55, no. 4, pp. 644–648, 1953. doi: 10.1042/bj0550644

[107] P. Schuck, "On the analysis of protein self-association by sedimentation velocity analytical ultracentrifugation," *Anal. Biochem.*, vol. 320, no. 1, pp. 104–124, 2003. doi: 10.1016/S0003-2697(03)00289-6

[108] H. Zhao, P.H. Brown, and P. Schuck, "On the distribution of protein refractive index increments," *Biophys. J.*, vol. 100, no. 9, pp. 2309–2317, 2011. doi: 10.1016/j.bpj.2011.03.004

[109] A.K. Attri, M.S. Lewis, and E.D. Korn, "The formation of actin oligomers studied by analytical ultracentrifugation." *J. Biol. Chem.*, vol. 266, no. 11, pp. 6815–24, 1991.

[110] R.L. Baldwin and J.W. Williams, "Boundary spreading in sedimentation velocity experiments," *J. Am. Chem. Soc.*, vol. 72, no. 9, p. 4325, 1950. doi: 10.1021/ja01165a554

[111] R.L. Baldwin, "Boundary spreading in sedimentation velocity experiments. VI. A better method for finding distributions of sedimentation coefficient when the effects of diffusion are large," *J. Phys. Chem.*, vol. 63, no. 10, pp. 1570–1573, 1959. doi: 10.1021/j150580a006

[112] J.W. Williams, R.L. Baldwin, M. Saunders, and P.G. Squire, "Boundary spreading in sedimentation velocity experiments. I. The enzymatic degradation of serum globulins," *J. Am. Chem. Soc.*, vol. 74, no. 6, pp. 1542–1548, 1952. doi: 10.1021/ja01126a059

[113] T. Svedberg and H. Rinde, "The ultra-centrifuge, a new instrument for the determination of size and distribution of size of particle in amicroscopic colloids," *J. Am. Chem. Soc.*, vol. 46, no. 1923, pp. 2677–2693, 1924. doi: 10.1021/ja01677a011

[114] R.L. Baldwin, "Boundary spreading in sedimentation velocity experiments. III. Effects of diffusion on the measurement of heterogeneity when concentration dependence is absent," *J. Phys. Chem.*, vol. 58, no. 2, pp. 1081–1086, 1954. doi: 10.1021/j150522a009

[115] P.J. Wan and E.T. Adams, "Molecular weights and molecular-weight distributions from ultracentrifugation of nonideal solutions," *Biophys. Chem.*, vol. 5, no. 1–2, pp. 207–241, 1976. doi: 10.1016/0301-4622(76)80036-1

[116] J.W. Williams, *Ultracentrifugation of Macromolecules.* New York: Academic Press, 1972.

[117] K.E. van Holde and W.O. Weischet, "Boundary analysis of sedimentation-velocity experiments with monodisperse and paucidisperse solutes," *Biopolymers*, vol. 17, no. 6, pp. 1387–1403, 1978. doi: 10.1002/bip.1978.360170602

[118] G.A. Gilbert and L.M. Gilbert, "Detection in the ultracentrifuge of protein heterogeneity by computer modelling, illustrated by pyruvate dehydrogenase multienzyme complex," *J. Mol. Biol.*, vol. 144, no. 3, pp. 405–408, 1980. doi: 10.1016/0022-2836(80)90099-6

[119] L.W. Nichol and D.J. Winzor, "Calculation of asymptotic boundary shapes from experimental mass migration patterns," *Methods Enzymol.*, vol. 130, no. 1968, pp. 6–18, 1986. doi: 10.1016/0076-6879(86)30004-1

[120] J. Behlke and O. Ristau, "Enhanced resolution of sedimentation coefficient distribution profiles by extrapolation to infinite time," *Eur. Biophys. J.*, vol. 39, no. 3, pp. 449–455, 2010. doi: 10.1007/s00249-009-0425-1

[121] P. Schuck and P Rossmanith, "Determination of the sedimentation coefficient distribution by least-squares boundary modeling," *Biopolymers*, vol. 54, no. 5, pp. 328–341, 2000. doi: 10.1002/1097-0282(20001015)54:5¡328::AID-BIP40¿3.0.CO;2-P

[122] S.W. Provencher, "CONTIN: A general purpose constrained regularization program for inverting noisy linear algebraic and integral equations." *Comput. Phys. Commun.*, vol. 27, pp. 229–242, 1982. doi: 10.1016/0010-4655(82)90174-6

[123] Y.W. Chiang, P.P. Borbat, and J.H. Freed, "Maximum entropy: A complement to Tikhonov regularization for determination of pair distance distributions by pulsed ESR." *Jmri-J. Magn. Reson. Im.*, vol. 177, no. 2, pp. 184–196, 2005. doi: 10.1016/j.jmr.2005.07.021

[124] S. Sibisi, J. Skilling, R.G. Brereton, E.D. Laue, and J. Staunton, "Maximum entropy signal processing in practical NMR spectroscopy," *Nature*, vol. 311, no. 5985, pp. 446–447, 1984. doi: 10.1038/311446a0

[125] A. Li, E.L. Miller, M.E. Kilmer, T.J. Brukilacchio, T. Chaves, J. Stott, Q. Zhang, T. Wu, M. Chorlton, R.H. Moore, D.B. Kopans, and D.A. Boas, "Tomographic optical breast imaging guided by three-dimensional mammography," *Appl. Opt.*, vol. 42, no. 25, pp. 5181–5190, 2003. doi: 10.1364/AO.42.005181

[126] R. Narayan and R. Nityananda, "Maximum entropy image restoration in astronomy," *Ann. Rev. Astron. Astrophys.*, vol. 24, pp. 127–170, 1986. doi: 10.1146/annurev.aa.24.090186.001015

[127] O. Coulon, D.C. Alexander, and S. Arridge, "Diffusion tensor magnetic resonance image regularization," *Med. Image Anal.*, vol. 8, no. 1, pp. 47–67, 2004. doi: 10.1016/j.media.2003.06.002

[128] J. Svitel, A. Balbo, R.A. Mariuzza, N.R. Gonzales, and P. Schuck, "Combined affinity and rate constant distributions of ligand populations from experimental surface-binding kinetics and equilibria." *Biophys. J.*, vol. 84, pp. 4062–4077, 2003. doi: 10.1016/S0006-3495(03)75132-7

[129] P.J. Steinbach, "Two-dimensional distributions of activation enthalpy and entropy from kinetics by the maximum entropy method," *Biophys. J.*, vol. 70, pp. 1521–1528, 1996.

[130] P.J. Steinbach, K. Chu, H. Frauenfelder, J.B. Johnson, D.C. Lamb, G.U. Nienhaus, T.B. Sauke, and R.D. Young, "Determination of rate distributions from kinetic experiments," *Biophys. J.*, vol. 61, pp. 235–245, 1992. doi: 10.1016/S0006-3495(92)81830-1

[131] P.J. Steinbach, R. Ionescu, and C.R. Matthews, "Analysis of kinetics using a hybrid maximum-entropy/nonlinear-least-squares method: Application to protein folding," *Biophys. J.*, vol. 82, no. 4, pp. 2244–2255, 2002. doi: 10.1016/S0006-3495(02)75570-7

[132] P.R. Bevington and D.K. Robinson, *Data Reduction and Error Analysis for the Physical Sciences.* New York: Mc-Graw-Hill, 1992.

[133] D.L. Phillips, "A technique for the numerical solution of certain integral equations of the first kind," *Assoc. Comput. Mach.*, vol. 9, pp. 84–97, 1962. doi: 10.1145/321105.321114

[134] P.C. Hansen, "Numerical tools for analysis and solution of Fredholm integral equations of the first kind," *Inverse Probl.*, vol. 8, no. 6, pp. 849–872, 1992. doi: 10.1088/0266-5611/8/6/005

[135] U. Amato and W. Hughes, "Maximum entropy regularization of Fredholm integral equations of the first kind," *Inverse Probl.*, vol. 7, no. 6, pp. 793–808, 1991. doi: 10.1088/0266-5611/7/6/004

[136] J. Skilling, "Maximum entropy image reconstruction-general algorithm," *Mon. Not. R. Astro. Soc.*, vol. 211, pp. 111–124, 1984. doi: 10.1093/mnras/211.1.111

[137] P. Schuck, "On computational approaches for size-and-shape distributions from sedimentation velocity analytical ultracentrifugation," *Eur. Biophys. J.*, vol. 39, no. 8, pp. 1261–1275, 2010. doi: 10.1007/s00249-009-0545-7

[138] S. Trachtenberg, P. Schuck, T.M. Phillips, S.B. Andrews, and R.D. Leapman, "A structural framework for a near-minimal form of life: Mass and compositional analysis of the helical mollicute *Spiroplasma melliferum* BC3," *PLoS One*, vol. 9, no. 2, p. e87921, 2014. doi: 10.1371/journal.pone.0087921

[139] Y.-F. Mok, G.J. Howlett, and M.D.W. Griffin, "Sedimentation velocity analysis of the size distribution of amyloid oligomers and fibrils," *Methods Enzymol.*, pp. 241–256, 2015. doi: 10.1016/bs.mie.2015.06.024

[140] N. Gupta, J. Arthos, P. Khazanie, T.D. Steenbeke, N.M. Censoplano, E.A. Chung, C.C. Cruz, M.A. Chaikin, M. Daucher, S. Kottilil, D. Mavilio, P. Schuck, P.D. Sun, R.L. Rabin, S. Radaev, D. Van Ryk, C. Cicala, and A.S. Fauci, "Targeted lysis of HIV-infected cells by natural killer cells armed and triggered by a recombinant immunoglobulin fusion protein: Implications for immunotherapy," *Virology*, vol. 332, no. 2, pp. 491–497, 2005. doi: 10.1016/j.virol.2004.12.018

[141] S.E. Harding, G.G. Adams, F. Almutairi, Q. Alzahrani, T. Erten, M.S. Kök, and R.B. Gillis, "Ultracentrifuge methods for the analysis of polysaccharides, glycoconjugates, and lignins," *Methods Enzymol.*, pp. 391–439, 2015. doi: 10.1016/bs.mie.2015.06.043

[142] V. Vogel, K. Langer, S. Balthasar, P. Schuck, W. Mächtle, W. Haase, J.A. van den Broek, C. Tziatzios, and D. Schubert, "Characterization of serum albumin nanoparticles by sedimentation velocity analysis and electron microscopy," *Prog.Colloid Polym. Sci*, vol. 119, pp. 31–36, 2002. doi: 10.1007/3-540-44672-9_5

[143] W. Mächtle and L Börger, *Analytical Ultracentrifugation of Polymers and Nanoparticles*. Berlin: Springer, 2006. ISBN 978-3-540-26218-3

[144] J. Walter, T. Thajudeen, S. Süß, D. Segets, and W. Peukert, "New possibilities of accurate particle characterisation by applying direct boundary models to analytical centrifugation," *Nanoscale*, vol. 7, no. 15, pp. 6574–6587, 2015. doi: 10.1039/C5NR00995B

[145] S.A. Berkowitz and J.S. Philo, "Monitoring the homogeneity of adenovirus preparations (a gene therapy delivery system) using analytical ultracentrifugation," *Anal. Biochem.*, vol. 362, pp. 16–37, 2006. doi: 10.1016/j.virol.2004.12.018

[146] W.B. Bridgman, "Some physical characteristics of glycogen," *J. Am. Chem. Soc.*, vol. 64, no. 10, pp. 2349–2356, 1942. doi: 10.1021/ja01262a037

[147] H.-J. Cantow, "Zur Bestimmung von Teilchengroessenverteilungen in der Ultrazentrifuge," *Makromol. Chem.*, vol. 70, pp. 130–149, 1964. doi: 10.1002/macp.1964.020700110

[148] H.G. Müller, "Automated determination of particle-size distributions of dispersions by analytical ultracentrifugation," *Colloid Polym. Sci.*, vol. 267, no. 12, pp. 1113–1116, 1989. doi: 10.1007/BF01496933

[149] B. Ortlepp and D. Panke, "Analytical ultracentrifuges with multiplexer and video systems for measuring particle size and molecular mass distribution," *Prog. Coll. Polym. Sci.*, vol. 86, pp. 57–61, 1991. doi: 10.1007/BFb0115007

[150] W.F. Stafford, "Boundary analysis in sedimentation transport experiments: A procedure for obtaining sedimentation coefficient distributions using the time derivative of the concentration profile," *Anal. Biochem.*, vol. 203, no. 2, pp. 295–301, 1992. doi: 10.1016/0003-2697(92)90316-Y

[151] P. Schuck, "Sedimentation coefficient distributions of large particles," *Analyst*, vol. in press, 2016. doi: 10.1039/C6AN00534A

[152] J.S. Philo, "Improved methods for fitting sedimentation coefficient distributions derived by time-derivative techniques," *Anal. Biochem.*, vol. 354, no. 2, pp. 238–246, 2006. doi: 10.1016/j.ab.2006.04.053

[153] J.S. Philo, "A method for directly fitting the time derivative of sedimentation velocity data and an alternative algorithm for calculating sedimentation coefficient distribution functions." *Anal. Biochem.*, vol. 279, no. 2, pp. 151–163, 2000. doi: 10.1006/abio.2000.4480

[154] P.H. Brown and P. Schuck, "Macromolecular size-and-shape distributions by sedimentation velocity analytical ultracentrifugation," *Biophys. J.*, vol. 90, no. 12, pp. 4651–4661, 2006. doi: 10.1529/biophysj.106.081372

[155] R.P. Carney, J.Y. Kim, H. Qian, R. Jin, H. Mehenni, F. Stellacci, and O.M. Bakr, "Determination of nanoparticle size distribution together with density or molecular weight by 2D analytical ultracentrifugation." *Nat. Commun.*, vol. 2, no. May, p. 335, jan 2011. doi: 10.1038/ncomms1338

[156] B. J. Berne and R. Pecora, *Dynamic Light Scattering.* Mineola, New York: Dover Publications, 2000.

[157] P. Schuck, "Measuring size-and-shape distributions of protein complexes in solution by sedimentation and dynamic light scattering," Paper presented at Euroconference Advances in Analytical Ultracentrifugation and Hydrodynamics, Autrans, France, 2002.

[158] J. Vistica, J. Dam, A. Balbo, E. Yikilmaz, R.A. Mariuzza, T.A. Rouault, and P. Schuck, "Sedimentation equilibrium analysis of protein interactions with global implicit mass conservation constraints and systematic noise decomposition," *Anal. Biochem.*, vol. 326, no. 2, pp. 234–256, 2004. doi: 10.1016/j.ab.2003.12.014

[159] S.E. Harding, P. Schuck, A.S. Abdelhameed, G. Adams, M.S. Kök, and G.A. Morris, "Extended Fujita approach to the molecular weight distribution of polysaccharides and other polymeric systems." *Methods*, vol. 54, no. 1, pp. 136–44, 2011. doi: 10.1016/j.ymeth.2011.01.009

[160] C. Ebel, "Sedimentation velocity to characterize surfactants and solubilized membrane proteins." *Methods*, vol. 54, no. 1, pp. 56–66, 2011. doi: 10.1016/j.ymeth.2010.11.003

[161] B. Demeler, T.-L. Nguyen, G.E. Gorbet, V. Schirf, E.H. Brookes, P. Mulvaney, A.O. El-Ballouli, J. Pan, O.M. Bakr, A.K. Demeler, B.I. Hernandez Uribe, N. Bhattarai, and R.L. Whetten, "Characterization of size, anisotropy, and density heterogeneity of nanoparticles by sedimentation velocity." *Anal. Chem.*, vol. 86, no. 15, pp. 7688–7695, 2014. doi: 10.1021/ac501722r

[162] C.A. MacRaild, D.M. Hatters, L.J. Lawrence, and G.J. Howlett, "Sedimentation velocity analysis of flexible macromolecules: Self-association and tangling of amyloid fibrils." *Biophys. J.*, vol. 84, no. 4, pp. 2562–2569, 2003. doi: 10.1016/S0006-3495(03)75061-9

[163] H. Yamakawa, "Statistical mechanics of wormlike chains," *Pure Appl. Chem.*, vol. 46, no. 2, pp. 135–141, 1976.

[164] H. Yamakawa and M. Fujii, "Translational friction coefficient of wormlike chains," *Macromolecules*, vol. 6, no. 3, pp. 407–415, 1973. doi: 10.1021/ma60033a018

[165] G.H. Koenderink, K.L. Planken, R. Roozendaal, and A.P. Philipse, "Monodisperse DNA restriction fragments II. Sedimentation velocity and equilibrium experiments," *J. Coll. Interf. Sci.*, vol. 291, no. 1, pp. 126–134, 2005. doi: 10.1016/j.jcis.2005.04.114

[166] B.H. Zimm, "Chain molecule hydrodynamics by the Monte-Carlo method and the validity of the Kirkwood-Riseman approximation," *Macromolecules*, vol. 13, no. 3, pp. 592–602, 1980. doi: 10.1021/ma60075a022

[167] A.A. Sousa and R.D. Leapman, "Mass mapping of amyloid fibrils in the electron microscope using STEM imaging," in *Nanoimaging.* Totowa, NJ: Humana Press, 2013, pp. 195–207.

[168] W.-F. Xue, S.W. Homans, and S.E. Radford, "Amyloid fibril length distribution quantified by atomic force microscopy single-particle image analysis," *Protein Eng. Des. Sel.*, vol. 22, no. 8, pp. 489–496, 2009. doi: 10.1093/protein/gzp026

[169] D.S. Sivia, *Data Analysis. A Bayesian Tutorial.* Oxford: Oxford University Press, 1996. ISBN 978-0198568322

[170] L. Wafer, M. Kloczewiak, and Y. Luo, "Quantifying trace amounts of aggregates in biopharmaceuticals using analytical ultracentrifugation sedimentation velocity: Bayesian analyses and F statistics," *AAPS J.*, no. 7, 2016. doi: 10.1208/s12248-016-9925-y

[171] B. Elzen, *Scientists and Rotors. The Development of Biochemical Ultracentrifuges.* Enschede: Dissertation, University Twente, 1988.

[172] R. Signer and H. Gross, "Ultrazentrifugale Polydispersitätsbestimmungen an hochpolymeren Stoffen," *Helv. Chim. Acta*, vol. 17, no. 1, pp. 726–735, 1934. doi: 10.1002/hlca.19340170188

[173] R.L. Baldwin, L.J. Gosting, J.W. Williams, and R.A. Alberty, "Characterization and physical properties. Transport processes and the heterogeneity of proteins," *Discuss. Faraday Soc.*, vol. 20, no. I, p. 13, 1955. doi: 10.1039/df9552000013

[174] J.W. Williams, K.E. van Holde, R.L. Baldwin, and H. Fujita, "The theory of sedimentation analysis," *Chem. Rev.*, vol. 58, no. 4, pp. 715–744, 1958. doi: 10.1021/cr50022a005

[175] L.J. Gosting, "Solution of boundary spreading equations for electrophoresis and the velocity ultracentrifuge," *J. Am. Chem. Soc.*, vol. 74, no. 1950, pp. 1548–1552, 1952. doi: 10.1021/ja01126a060

[176] D.J. Winzor, R. Tellam, and L.W. Nichol, "Determination of the asymptotic shapes of sedimentation velocity patterns for reversibly polymerizing solutes," *Arch. Biochem. Biphys.*, vol. 178, no. 2, pp. 327–322, 1977. doi: 10.1016/0003-9861(77)90200-4

[177] C.A. Brautigam, "Calculations and publication-quality illustrations for analytical ultracentrifugation data," *Methods Enzymol.*, vol. 562, pp. 109–133, 2015. doi: 10.1016/bs.mie.2015.05.001

[178] N. Gralén and G. Lagermalm, "A contribution to the knowledge of some physicochemical properties of polystyrene," *J. Phys. Chem.*, vol. 56, pp. 514–523, 1952. doi: 10.1021/j150496a025

[179] B. Demeler and K.E. van Holde, "Sedimentation velocity analysis of highly heterogeneous systems," *Anal. Biochem.*, vol. 335, no. 2, pp. 279–288, 2004. doi: 10.1016/j.ab.2004.08.039

[180] R.L. Baldwin, "Boundary spreading in sedimentation velocity experiments. II. The correction of sedimentation coefficient distributions for the dependence of sedimentation coefficient on concentration," *J. Am. Chem. Soc.*, vol. 76, no. 2, pp. 402–407, 1954. doi: 10.1021/ja01631a026

[181] R.L. Baldwin, "Boundary spreading in sedimentation-velocity experiments. 4. Measurement of the standard deviation of a sedimentation-coefficient distribution: Application to bovine albumin and β-lactoglobulin," *Biochem. J.*, vol. 65, no. 1953, pp. 490–502, 1956. doi: 10.1042/bj0650490

[182] B. Demeler, H. Saber, and J.C. Hansen, "Identification and interpretation of complexity in sedimentation velocity boundaries," *Biophys. J.*, vol. 72, no. 1, pp. 397–407, 1997. doi: 10.1016/S0006-3495(97)78680-6

[183] W.F. Stafford, "Boundary analysis in sedimentation velocity experiments," *Methods Enzymol.*, vol. 240, pp. 478–501, 1994. doi: 10.1016/S0076-6879(94)40061-X

[184] R. Cohen, J. Cluzel, H. Cohen, P. Male, M. Moignier, and C. Soulie, "MaD, An automated precise analytical ultracentrifuge scanner system," *Biophys. Chem.*, vol. 5, no. 1–2, pp. 77–96, 1976. doi: 10.1016/0301-4622(76)80027-0

[185] W.F. Stafford, "Methods for obtaining sedimentation coefficient distributions," in *Anal. Ultracentrifugation Biochem. Polym. Sci.*, S.E. Harding, A.J. Rowe, and J.C. Horton, Eds. Cambridge, U.K.: The Royal Society of Chemistry, 1992, pp. 359–393.

[186] B. Kokona, C.A. May, N.R. Cunningham, L. Richmond, F.J. Garcia, J.C. Durante, K.M. Ulrich, C.M. Roberts, C.D. Link, W.F. Stafford, T.M. Laue, and R. Fairman, "Studying polyglutamine aggregation in *Caenorhabditis elegans* using an analytical ultracentrifuge equipped with fluorescence detection." *Protein Sci.*, vol. 25, pp. 1–13, 2015. doi: 10.1002/pro.2854

[187] D.A. Yphantis, J.W. Lary, W.F. Stafford, S. Liu, P.H. Olsen, D.B. Hayes, T.P. Moody, T.M. Ridgeway, D.A. Lyons, and T.M. Laue, "On-line data acquisition for the Rayleigh interference optical system of the analytical ultracentrifuge," in *Modern Analytical Ultracentrifugation: Techniques and Methods*, T.M. Schuster and T.M. Laue, Eds. Boston: Birkhäuser, 1994, pp. 209–226.

[188] T.M. Laue, L.A. Anderson, and P.D. Demaine, "An on-line interferometer for the XL-A ultracentrifuge," *Prog. Coll. Polym. Sci.*, vol. 94, pp. 74–81, 1994. doi: 10.1007/BFb0115604

[189] W.F. Stafford, "Sedimentation velocity spins a new weave for an old fabric," *Curr. Opin. Biotechnol.*, vol. 8, no. 1, pp. 14–24, 1997. doi: 10.1016/S0958-1669(97)80152-8

[190] J.M. Beechem, "Global analysis of biochemical and biophysical data," *Methods Enzymol.*, vol. 210, pp. 37–54, 1992. doi: 1016/0076-6879(92)10004

[191] E. Brookes, W. Cao, and B. Demeler, "A two-dimensional spectrum analysis for sedimentation velocity experiments of mixtures with heterogeneity in molecular weight and shape." *Eur. Biophys. J.*, vol. 39, no. 3, pp. 405–14, 2010. doi: 10.1007/s00249-009-0413-5

[192] J. Pearson, F. Krause, D. Haffke, B. Demeler, K. Schilling, and H. Cölfen, "Next-generation AUC adds a spectral dimension: Development of multiwavelength detectors for the analytical ultracentrifuge," *Methods Enzymol.*, vol. 562, pp. 1–26, 2015. doi: 10.1016/bs.mie.2015.06.033

[193] J. Walter, P.J. Sherwood, W. Lin, D. Segets, W.F. Stafford, and W. Peukert, "Simultaneous analysis of hydrodynamic and optical properties using analytical ultracentrifugation equipped with multiwavelength detection," *Anal. Chem.*, vol. 87, pp. 3396–3403, 2015. doi: 10.1021/ac504649c

[194] J.C.D. Houtman, H. Yamaguchi, M. Barda-Saad, A. Braiman, B. Bowden, E. Appella, P. Schuck, and L.E. Samelson, "Oligomerization of signaling complexes by the multipoint binding of GRB2 to both LAT and SOS1," *Nat. Struct. Mol. Biol.*, vol. 13, no. 9, pp. 798–805, 2006. doi: 10.1038/nsmb1133

[195] M. Barda-Saad, N. Shirasu, M.H. Pauker, N. Hassan, O. Perl, A. Balbo, H. Yamaguchi, J.C.D. Houtman, E. Appella, P. Schuck, and L.E. Samelson, "Cooperative interactions at the SLP-76 complex are critical for actin polymerization." *EMBO J.*, vol. 29, no. 14, pp. 2315–2328, 2010. doi: 10.1038/emboj.2010.133

[196] N.P. Coussens, R. Hayashi, P.H. Brown, L. Balagopalan, A. Balbo, I. Akpan, J.C.D. Houtman, V.A. Barr, P. Schuck, E. Appella, and L.E. Samelson, "Multipoint binding of the SLP-76 SH2 domain to ADAP is critical for oligomerization of SLP-76 signaling complexes in stimulated T cells." *Mol. Cell. Biol.*, vol. 33, no. 21, pp. 4140–4151, 2013. doi: 10.1128/MCB.00410-13

[197] S.B. Padrick and C.A. Brautigam, "Evaluating the stoichiometry of macromolecular complexes using multisignal sedimentation velocity." *Methods*, vol. 54, no. 1, pp. 39–55, 2011. doi: 10.1016/j.ymeth.2011.01.002

[198] S.B. Padrick, R.K. Deka, J.L. Chuang, R.M. Wynn, D.T. Chuang, M.V. Norgard, M.K. Rosen, and C.A. Brautigam, "Determination of protein complex stoichiometry through multisignal sedimentation velocity experiments," *Anal. Biochem.*, vol. 407, no. 1, pp. 89–103, 2010. doi: 10.1016/j.ab.2010.07.017

[199] H. Zhao, P.H. Brown, M.T. Magone, and P. Schuck, "The molecular refractive function of lens γ-crystallins," *J. Mol. Biol.*, vol. 411, no. 3, pp. 680–699, 2011. doi: 10.1016/j.jmb.2011.06.007

[200] H. Zhao, S. Lomash, C. Glasser, M.L. Mayer, and P. Schuck, "Analysis of high affinity self-association by fluorescence optical sedimentation velocity analytical ultracentrifugation of labeled proteins: Opportunities and limitations," *PLoS One*, vol. 8, no. 12, p. e83439, 2013. doi: 10.1371/journal.pone.0083439

[201] M. Melikishvili, D.W. Rodgers, and M.G. Fried, "6-Carboxyfluorescein and structurally similar molecules inhibit DNA binding and repair by O(6)-alkylguanine DNA alkyltransferase." *DNA Repair*, vol. 10, no. 12, pp. 1193–1202, 2011. doi: 10.1016/j.dnarep.2011.09.007

[202] P. Schuck, "Sedimentation velocity in the study of reversible multiprotein complexes," in *Protein Interactions: Biophysical Approaches for the study of complex reversible systems*, P. Schuck, Ed. New York: Springer, 2007, pp. 469–518.

[203] H. Zhao, M.L. Mayer, and P. Schuck, "Analysis of protein interactions with picomolar binding affinity by fluorescence-detected sedimentation velocity," *Anal. Chem.*, vol. 18, no. 6, pp. 3181–3187, 2014. doi: 10.1021/ac500093m

[204] H. Zhao, A.J. Berger, P.H. Brown, J. Kumar, A. Balbo, C.A. May, E. Casillas, T.M. Laue, G.H. Patterson, M.L. Mayer, and P. Schuck, "Analysis of high-affinity assembly for AMPA receptor amino-terminal domains." *J. Gen. Physiol.*, vol. 139, no. 5, pp. 371–388, 2012. doi: 10.1085/jgp.201210770

[205] R. Trautman, "Optical fine-structure of a meniscus in analytical ultracentrifugation in relation to molecular-weight determinations using the Archibald principle," *Biochim. Biophys. Acta*, vol. 28, pp. 417–431, 1958. doi: 10.1016/0006-3002(58)90490-6

[206] T.M.D. Besong, S.E. Harding, and D.J. Winzor, "The effective time of centrifugation for the analysis of boundary spreading in sedimentation velocity experiments." *Anal. Biochem.*, vol. 421, no. 2, pp. 755–758, 2012. doi: 10.1016/j.ab.2011.11.035

[207] P.F. Mijnlieff, P. van Es, and W.J.M. Jaspers, "Temperature gradients in ultracentrifuge cells due to adiabatic volume changes," *Recueil*, vol. 88, no. 2, pp. 220–224, 1969. doi: 10.1002/recl.19690880213

[208] M.A. Perugini, P. Schuck, and G.J. Howlett, "Differences in the binding capacity of human apolipoprotein E3 and E4 to size-fractionated lipid emulsions," *Eur J. Biochem.*, vol. 269, no. 23, pp. 5939–5949, 2002. doi: 10.1046/j.1432-1033.2002.03319.x

[209] P. Schuck, "Diffusion-deconvoluted sedimentation coefficient distributions for the analysis of interacting and non-interacting protein mixtures." in *Modern Analytical Ultracentrifugation: Techniques and Methods*, D.J. Scott, S.E. Harding, and A.J. Rowe, Eds. Cambridge: The Royal Society of Chemistry, 2006, pp. 26–50.

[210] S.L. Nyeo and B. Chu, "Maximum-entropy analysis of photon correlation spectroscopy data," *Macromolecules*, vol. 22, no. 10, pp. 3998–4009, 1989. doi: 10.1021/ma00200a031

[211] S.W. Provencher, "Low-bias macroscopic analysis of polydispersity," in *Laser Light Scatt. Biochem.*, S.E. Harding, D.B. Satelle, and V.A. Bloomfield, Eds. Cambridge, U.K.: The Royal Society of Chemistry, 1992, pp. 92–111.

[212] E.T. Jaynes, "Information theory and statistical mechanics," *Phys. Rev.*, vol. 106, no. 4, pp. 620–630, 1957. doi: 10.1103/PhysRev.106.620

[213] G.A. Gilbert and R.C.L. Jenkins, "Boundary problems in the sedimentation and electrophoresis of complex systems in rapid reversible equilibrium," *Nature*, vol. 177, no. 4514, pp. 853–854, 1956. doi: 10.1038/177853a0

[214] S.A. Berkowitz, "Role of analytical ultracentrifugation in assessing the aggregation of protein biopharmaceuticals," *AAPS J.*, vol. 8, no. 3, pp. E590–605, 2006. doi: 10.1208/aapsj080368

[215] J. Liu, J.D. Andya, and S.J. Shire, "A critical review of analytical ultracentrifugation and field flow fractionation methods for measuring protein aggregation," *AAPS J.*, vol. 8, no. 3, pp. E580–9, 2006. doi: 10.1208/aapsj080367

[216] A.H. Pekar and M. Sukumar, "Quantitation of aggregates in therapeutic proteins using sedimentation velocity analytical ultracentrifugation: Practical considerations that affect precision and accuracy," *Anal. Biochem.*, vol. 367, no. 2, pp. 225–237, 2007. doi: 10.1016/j.ab.2007.04.035

[217] J.P. Gabrielson and K.K. Arthur, "Measuring low levels of protein aggregation by sedimentation velocity." *Methods*, vol. 54, no. 1, pp. 83–91, 2011. doi: 10.1016/j.ymeth.2010.12.030

[218] S.A. Berkowitz and J.S. Philo, "Characterizing biopharmaceuticals using analytical ultracentrifugation," in *Biophys. Charact. Proteins Dev. Biopharm.*, D.J. Houde and S.A. Berkowitz, Eds. Amsterdam: Elsevier, 2015, ch. 9, pp. 211–260.

[219] G.M. Pavlov, D. Amoros, C. Ott, I.I. Zaitseva, J. Garcia de la Torre, and U.S. Schubert, "Hydrodynamic analysis of well-defined flexible linear macromolecules of low molar mass," *Macromolecules*, vol. 42, no. 19, pp. 7447–7455, 2009. doi: 10.1021/ma901027u

[220] G.M. Pavlov, K. Knop, O.V. Okatova, and U.S. Schubert, "Star-brush-shaped macromolecules: Peculiar properties in dilute solution," *Macromolecules*, vol. 46, no. 21, pp. 8671–8679, 2013. doi: 10.1021/ma400160f

[221] C.T. Chaton and A.B. Herr, "Elucidating complicated assembling systems in biology using size-and-shape analysis of sedimentation velocity data," *Methods Enzymol.*, vol. 562, pp. 187–204, 2015. doi: 10.1016/bs.mie.2015.04.004

[222] D.J. Cox, "Computer simulation of sedimentation in the ultracentrifuge. IV. Velocity sedimentation of self-associating solutes," *Arch. Biochem. Biophys.*, vol. 129, no. 1, pp. 106–123, 1969. doi: 10.1016/0003-9861(69)90157-X

[223] J.M. Claverie, H. Dreux, and R. Cohen, "Sedimentation of generalized systems of interacting particles. I. Solution of systems of complete Lamm equations," *Biopolymers*, vol. 14, no. 8, pp. 1685–1700, 1975. doi: 10.1002/bip.1975.360140811

[224] R.C. Chatelier, "A parameterized overspeeding method for the rapid attainment of low-speed sedimentation equilibrium," *Anal. Biochem.*, vol. 175, no. 1, pp. 114–119, 1988. doi: 10.1016/0003-2697(88)90368-5

[225] A.P. Minton, "Simulation of the time course of macromolecular separations in an ultracentrifuge. I. Formation of a cesium chloride density gradient at 25 degrees C," *Biophys. Chem.*, vol. 42, no. 1, pp. 13–21, 1992. doi: 10.1016/0301-4622(92)80004-O

[226] W. Cao and B. Demeler, "Modeling analytical ultracentrifugation experiments with an adaptive space-time finite element solution of the Lamm equation," *Biophys. J.*, vol. 89, no. 3, pp. 1589–1602, 2005. doi: 10.1529/biophysj.105.061135

[227] D.J. Cox, "Computer simulation of sedimentation in the ultracentrifuge II. Concentration-independent sedimentation," *Arch. Biochem. Biophys.*, vol. 112, no. 2, pp. 259–266, 1965. doi: 10.1016/0003-9861(65)90044-5

[228] G.H. Golub and C.F. van Loan, *Matrix Computations*. Baltimore, MD: The Johns Hopkins University Press, 1989. ISBN 0801854148

[229] C.L. Lawson and R.J. Hanson, *Solving Least Squares Problems*. Englewood Cliffs, New Jersey: Prentice-Hall, 1974. ISBN 0898713560

Index

Printed and bound by CPI Group (UK) Ltd, Croydon, CR0 4YY

01/11/2024

01782603-0005